Game of Rhombs / Game of Triangles

Game of Rhombs / Game of Triangles

Raymond Fletcher III

Printed in the United States of America.

Library of Congress Control Number: 2022941441

ISBN	Paperback	978-1-68536-660-5
	Hardback	978-1-68536-822-7
	eBook	978-1-68536-661-2

Westwood Books Publishing LLC
Atlanta Financial Center
3343 Peachtree Rd NE Ste 145-725
Atlanta, GA 30326

www.westwoodbookspublishing.com

CONTENTS

PREFACE

A domino like game which uses rhombs as playing tiles instead of dominoes provided inspiration for our subject. In the "Game of Rhombs", rhombs, (i.e., rhombuses), with interior angles of 60° and 120° are used to form various scoring configurations chiefly consisting of the simple hexagon, (H_1), and the *star*, (J_1) consisting of 6 rhombs with a common vertex. In the Game each tile is labeled with a number in the range {0, ..., 9}, and in a scoring configuration the tile values must sum to a multiple of 5. A complete set of rules for the Game of Rhombs is provided in Chapter 9.

A polyrhomb P is a polygon tiled with rhombs. The set of tiles incident with an interior vertex x of P is the neighborhood of x, (nbd(x)). If R denotes the set of tiles used to tile P, and G is an abelian group, then the mapping ϕ: R→G with the property that $\sum\{\phi(t) : t \in nbd(x)\} = 0$ for each interior vertex x of P, is a *locally nilpotent assignment from G.* If for every such locally nilpotent assignment, $\sum\{\{\phi(t) : t \in R\} = 0$, then we call P *nilpotent*. In the Game, the abelian group Z_5 is used as the base group and scoring configurations consist of nilpotent polyrhombs. For the general theory, the set Z of integers will be used exclusively as the base group.

The main purpose of the general theory will be to determine the nilpotence or nonnilpotence of various polygons. In Chapter 1 nilpotent hexagons composed of 60° rhombs like the ones occurring in the Game, are determined. A *semiregular polygon* is a convex polygon with opposite sides parallel and equal. In Chapter 2 the results of Chapter 1 are generalized to the category of semiregular polygons tiled with rhombs. The most general setting for the study of nilpotent figures is afforded by plane graphs. These are studied briefly in Chapter 3 with emphasis on plane graphs with rotational symmetry.

In Chapter 4 the problem of constructing locally nilpotent assignments for a given plane graph is considered, especially concerning graphs determined by tilings with rhombs. In Chapter 5 a special polyrhomb called a *flower* is introduced. A *free generating set, (FGS),* for a polyrhomb P, is a set of tiles in P whose integer values can be chosen freely with the result that a uniquely determined nilpotent assignment is created. In Chapter 4 diametric paths are used to construct an FGS for a given polyrhomb. These are called *simple* since the generation of a locally nilpotent assignment is obviously accomplished. The flower polyrhomb provides examples of FGS's which are termed *obscure* since it is not immediately clear how any further values for a locally nilpotent assignment can be obtained. A flower consists of *rings* of tiles arranged around a central star. In Chapter 5 it is determined exactly which of these rings form obscure FGS's.

If P is a rectangular polysquare with a locally nilpotent assignment and R is a subrectangle of P, then the corner values of R are related by simple linear equations. In Chapter 6, these *corner equations* are used to obtain a characterization of FGS's in a rectangular polysquare. In Chapter 7 *stream polygons* in a polyrhomb are introduced as the analog of subrectangles in a polysquare. Corner equations of a similar nature are obtained and then used to obtain a characterization of FGS's in semiregular polyrhombs.

A *stream* in a polyrhomb P is a sequence $(t_1, t_2, \dots, t_k)$ of adjacent rhombs in P with all common sides parallel. The streams in P yield a partition of the edge set of the graph of P, and each tile in P occurs at the intersection of exactly two streams. In Chapter 8 a set Q_1 of nonparallelogram lattice quadrilaterals is introduced. If P is a polygon with a Q_1 tiling then a stream in P can also be defined: a sequence $(t_1, t_2, \dots, t_k)$ of adjacent Q_1 tiles with the property that the shared edge between t_i and t_{i+1} and the shared edge between t_{i+1} and t_{i+2} are opposite edges of the quadrilateral t_{i+1}. These streams also yield a partition of the edge set of P, and it remains true that every tile of P lies at the intersection of two streams. Nevertheless streams in a Q_1 tiling exhibit some properties that are not found in tilings with rhombs or parallelograms. For example, the same two streams can intersect each other in more than one tile.

Even with these differences, equations involving corners of stream polygons are obtained and ultimately used to obtain a characterization of nilpotent semiregular polygons tiled with convex quadrilaterals.

Chapter 9 gives the rules for the Game of Rhombs. The interested reader can easily construct the game tiles using mat board, an exacto knife, a straightedge and a compass. The Game has proved very excellent after testing it many times. Strategic thinking, circumspection, mental arithmetic and geometric recognition are all qualities inherent in the play.

Chapter 10 begins what might well be regarded as a second book in our manuscript. The purpose of identifying nilpotent polygons remains the same but takes on a very different character when equilateral (unit) triangles are used as tiles instead of rhombs or general quadrilaterals. The main difference in character is due to the fact that given a polygon there is one and only one way to tile it with unit triangles, whereas with rhombs many distinct tilings are possible. Also streams which are useful for the analysis of polyrhombs are nonexistent in polygons tiled with unit triangles, (we call these *polytriangles,* although they are referred to in [5] as *polyiamonds.*)

For a polytriangle P a locally nilpotent assignment from Z is defined as follows. The six unit triangles incident with a common vertex forming a simple hexagon H_1 and the triangle Δ_2 consisting of four unit triangles, are the basic figures which are required to have a 0 value in any locally nilpotent assignment. A polytriangle P is *nilpotent* in case for any locally nilpotent assignment, the *value of P*, (i.e., the sum of the values assigned to all unit triangles which constitute P, is 0. In Chapter 10 nilpotent triangulated triangles, trapezoids, parallelograms, pentagons and hexagons are determined. A polytriangle is *prime* in case it is nilpotent and has no dissection into smaller nilpotent polytriangles. Ultimately it is shown that H_1 and the semiregular hexagon with sides of length (1,2,2,1,2,2) are the only prime triangulated convex hexagons.

In Chapter 11 *triangulated stars* are examined. Let Δ_m denote the triangle with m edges per side tiled with unit triangles, and let H_m denote the triangulated regular hexagon with m edges per side. The *regular star* J_m is formed by attaching six copies of Δ_m to the sides of H_m. It is shown that J_m is composite, hence nilpotent when m>1. In case m, p are odd with p<m, and we attach a Δ_p to the center of each side of H_m an *irregular star* $[H_m : \Delta_p]$ is obtained. The irregular star $[H_3 : \Delta_1]$ is prime, but for m≥5 it is shown that $[H_m : \Delta_p]$ is a composite extension of various odd shaped prime polytriangles.

Finally in Chapter 12 rules are supplied for the "Game of Triangles". This game is similar to the "Game of Rhombs" but is played with equilateral triangles instead of rhombs. The main scoring configurations consist of the prime triangulated polygons Δ_2, H_1, J_1, and the semiregular hexagon H(1,2,2,1,2,2). Alternatively players may also consider Δ_4 and H_2 as additional scoring configurations.

Each Chapter, except for the Game descriptions in Chapters 9, 12, is provided with an Exercise Section which the reader can use to aid his/her understanding of the material and which also provides suggestions for further research. Special thanks go to my wife Sylvie with whom I could test both the Game of Rhombs and the Game of Triangles and refine the rules for each.

CHAPTER 1

Nilpotent Hexagons

1.1 Introduction

Our subject was suggested by the nature of scoring configurations in the *Game of Rhombs*. This domino like game is played with tiles in the shape of rhombuses, (i.e., *rhombs*), with interior angles 60°, 120°. Each tile is given a number in the range $\{0, \ldots, 9\}$ and scoring configurations consist of hexagons or star polygons the sum of whose constituent tiles is a multiple of 5. Another scoring configuration consists of a star embedded in a hexagon as in Figure 1.1.

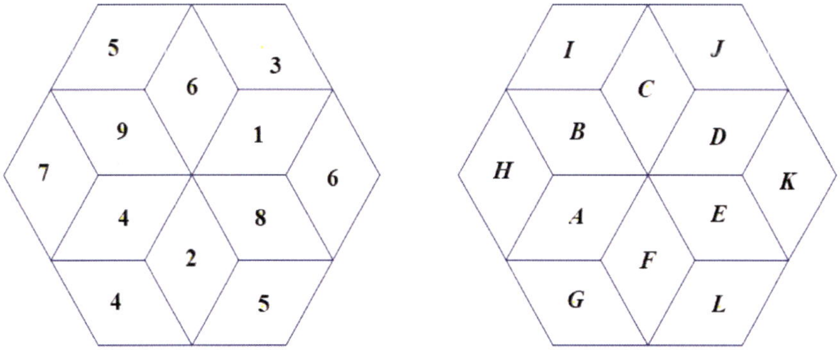

Figure 1.1

The interesting fact about this configuration is that if the central star and the six small hexagons are scoring configurations, then the entire large hexagon *must* also be a scoring configuration. This is easily proved: label the tile values as in the right hand side of Figure 1.1. As disjoint hexagons (F,E,L), (C,D,J), (A,B,H) are scoring configurations, the sum of their values is 0 (mod 5). So it suffices to show G+I+K = 0 (mod 5). Hexagons {G,A,F), (I,B,C), (K,D,E) are scoring configurations so we have

(1) G+A+F = 0 (mod 5)
(2) I+B+C = 0 (mod 5)
(3) K+D+E = 0 (mod 5).

Summing (1), (2), (3) yields

(4) (G+I+K) + (A+B+C+D+E+F) = 0 (mod 5),

and since the central star is a scoring configuration, we have (A+B+C+D+E+F) = 0 (mod 5), when (4) yields G+I+K = 0 (mod 5).

The following general definitions are suggested by the above example. A convex polygon P is *semiregular* in case opposite sides are parallel and equal. The boundary of P will be composed of edges which we will describe as vectors from a basic set V. This set consists of nonzero vectors all with different directions and with the property that if v∈V, then -v∈V. P has a *tiling by parallelograms* in case P is tiled entirely by parallelograms whose sides are composed of basic edges from V. In [1] it is shown that such a tiling can always be constructed. For our present purposes we will require that every vector in V have the same *(unit)* length, and thus all tiles will be rhombs. In this case P is called a *polyrhomb.* In Figure 1.2 is illustrated a typical semiregular polygon tiled with rhombs.

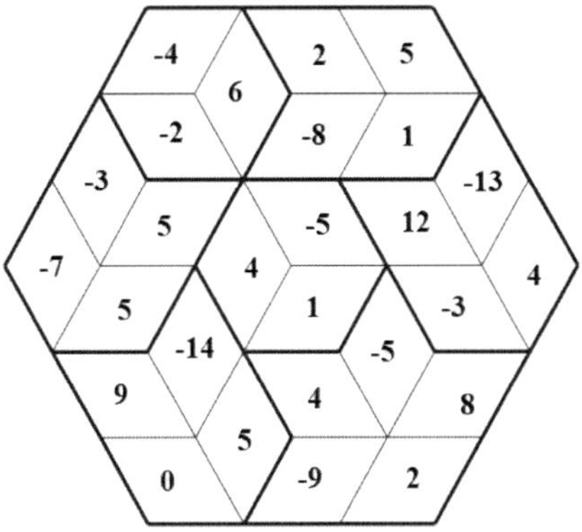

Figure 1.2

Note that elements from the abelian group Z of integers have been assigned to each tile in Figure 1.2 so that the *neighborhood (nbd)* of each interior vertex has sum 0. We say that the resulting polygon has a *locally nilpotent assignment*. Observe also that the bold outlines indicate a dissection of the hexagon into nbds. A polygon P tiled with parallelograms is *globally nilpotent, (or just nilpotent)* in case every locally nilpotent assignment from any abelian group results in P being globally nilpotent. Clearly any polygon with a nbd dissection is nilpotent.

1.2 Hexagons with all sides of odd length

The notation H(p,q,r) will be used to denote a hexagon with semiperimeter consisting of sides with length p, q, r. If p,q,r are all odd, then any tiling of H = H(p,q,r) by rhombs has the following nice property.

Theorem 1.1: *If p,q,r are odd, then any tiling of H(p,q,r) has a unique nbd dissection.*

Proof. Consider the example in Figure 1.3. Let H = H(p,q,r) be a semiregular hexagon with a tiling T by rhombs. Note that if Δ is a tiling of H with unit equilateral triangles then every vertex of Δ would also be a vertex of T. Thus every interior vertex of T is the center point of a nbd. A set S of interior points of T can be selected which form a triangular lattice in which adjacent points are two units apart. These points are indicated in the example in Figure 1.3.

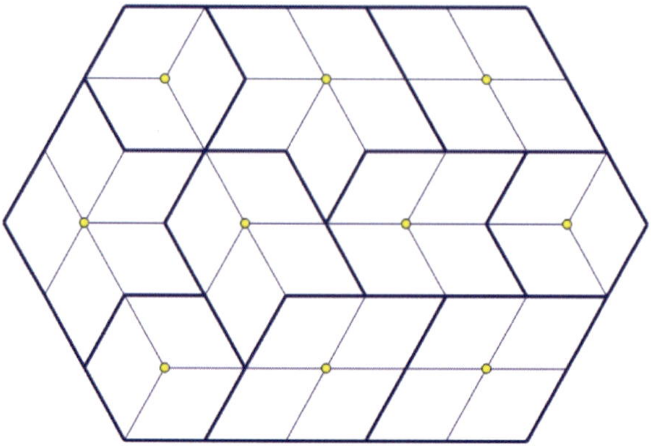

Figure 1.3

For each point x in S let N(x) denote the corresponding nbd. We claim that these nbds provide a dissection of H. First note that no tile can be incident with more than one point in S since the points in S are 2 units apart and the long diagonal of each rhomb is √(3). In Figure 1.4 a tiling of H with unit equilateral triangles and hexagons is illustrated. The points in S serve as centers of the hexagons. From this diagram it is clear that every rhomb in the tiling T must be incident with one of the points in S. As a consequence {N(x): x ∈S} provides a nbd dissection of H.

That this nbd dissection is unique can be easily seen from Figure 1.5 where for each of the points p_i in S a corresponding tile T_i is indicated. The tiles T_i are easily found with the property that they can belong only to the neighborhood $N(p_i)$. □

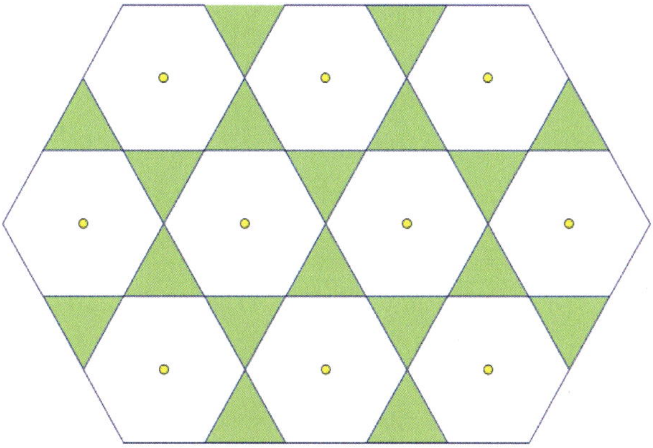

Figure 1.4

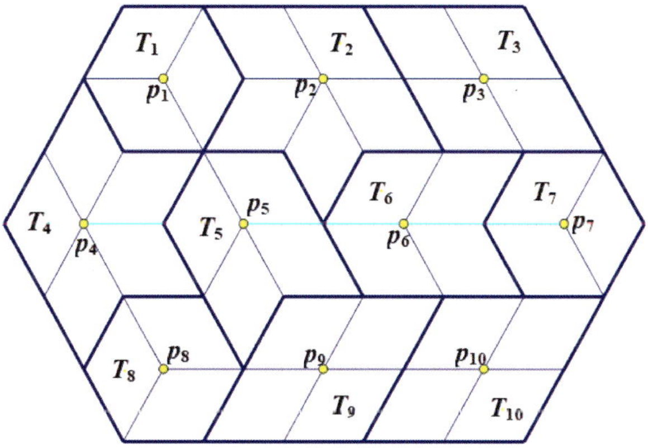

Figure 1.5

There are 5 kinds of nbds which occur in our dissection. These can all be found in the H_5 = H(5,5,5) hexagon in Figure 1.6. We will refer to this group of tiles as R(60) meaning all nbds which can be created using rhombs with interior angles a multiple of 60 degrees. In general, if angle α is a divisor of 360° then R(α) will denote all nbds which can be formed using rhombs with all interior angles a multiple of α. The tiles in R(60) will be labeled as in Figure 1.6: S for *star*; C for *chevron*; P for *parallelogram*; A for *arrow*, and H_1 for *hexagon*. The *composition* of an R(60) tiling represents the number of each type of tile which appears. Thus the composition of the tiling in Figure 1.6 is $H_1^5C^4P^7A^2S^1$.

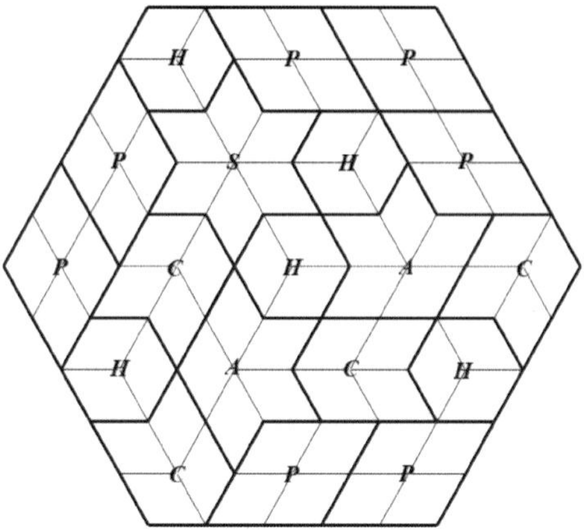

Figure 1.6

In Section 1.5 some puzzles involving R(60) tiles are given. An interesting problem regarding R(60) tiles involves *inversions.* An inversion can occur when two adjacent R(60) tiles have two common edges. For example, the adjacent H_1 and A tiles can be *inverted* or retiled by two C tiles as illustrated in Figure 1.7. An adjacent C and H_1 tile can be tiled differently with C and H_1. Also indicated in Figure 1.7, an adjacent S and H_1 tile can be retiled using an A and a C tile. An adjacent S and P tile can be retiled using two A tiles. Every hexagon with odd length sides can be tiled as in Figure 1.8 using only C tiles and one H_1 tile. Call this tiling H_1C^m. Now we make an interesting Conjecture.

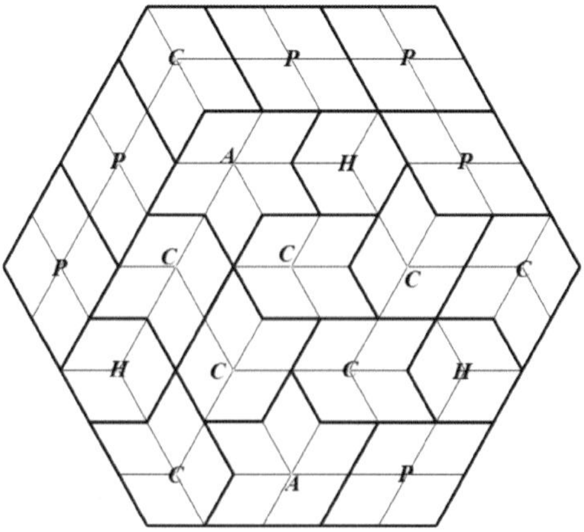

Figure 1.7

Conjecture 1.2 By means of a series of R(60) inversions as described above, it is possible to transform any R(60) tiling of a semiregular hexagon into an H_1C^m tiling.

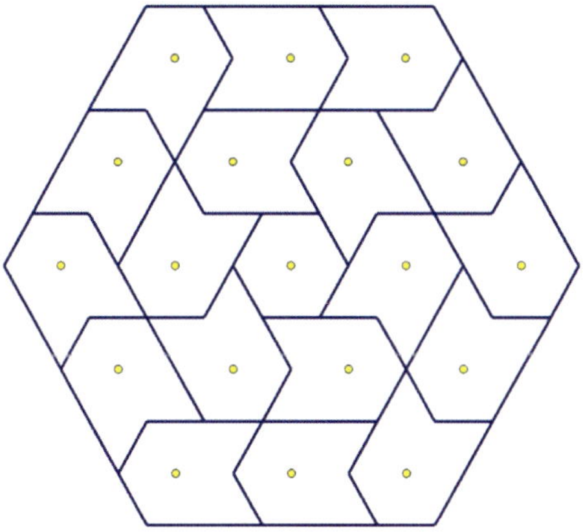

Figure 1.8

As evidence for Conjecture 1.2 consider the series of H_3 = H(3,3,3) hexagons in Figure 1.9. Starting with the R(60) tiling in 1.9(I) the transformation to an HC^6 tiling in 1.9(X) is accomplished by the series of indicated inversions.

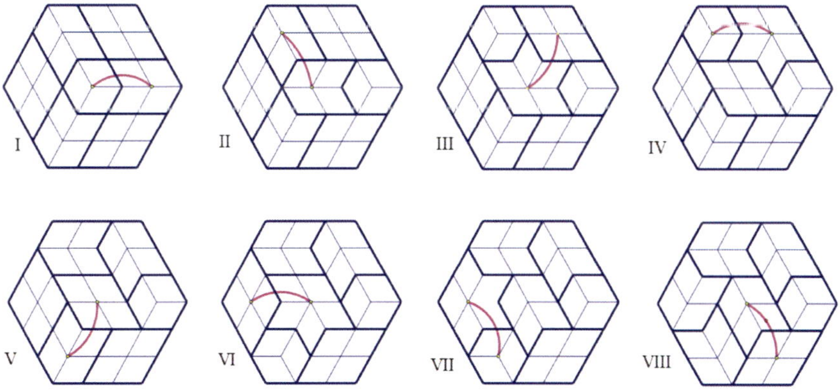

Figure 1.9

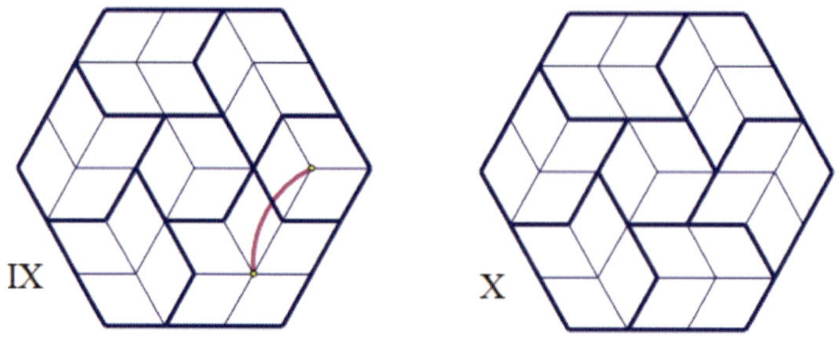

Figure 1.9

1.3 Hexagons with all sides of even length

Hexagons with all sides of even length require a very different analysis. The following Theorem is of quite a general nature as it applies to any nilpotent semiregular polygon. Of key importance is a simple operation involving the hexagon $H_{1.}$ Note that H_1 can be tiled in two different ways. The process of changing from one of these tilings to the other will be called an *H_1-inversion.*

Theorem 1.3 *H_1-inversion preserves nilpotence.*

Proof. In Figure 1.10(I) a section of a nilpotent semiregular polygon P which contains an H_1 is illustrated. The three tiles which constitute H_1 are labeled with values A, B, C from the base group. In Figure 1.10(II) a section of a polygon P′ is illustrated which is identical with P except that the H_1 in P has been inverted. The inverted hexagon is labeled with values $A' = A+B$; $B' = C+B$, and $C' = A+C$. In both Figures the vertices of H_1 are labeled {p,q,r,s,t,u}. Since P is assumed to be nilpotent we have $\Sigma N(x) = 0$ for each point $x \in \{p,q,r,s,t,u\}$. The corresponding nbds in P′ will be denoted $N'(x)$, and by our choice of A′, B′, C′ we at least have $\Sigma N'(x) = 0$ for $x \in \{q,s,u\}$.

To show that P′ is nilpotent, it suffices to show that $A'+ B'+ C' = 0$ and that $\Sigma N'(x) = 0$ for $x \in \{p,r,t\}$. Since P is nilpotent we have $A+B+C=0$ and thus $A'+ B'+ C' = 2(A+B+C)=0$. Consider: $\Sigma N'(p) = A'+C'+\Sigma N(p) - A = (A+B) + (A+C) + 0 - A = A+B+C = 0$. Similarly $\Sigma N'(r) = \Sigma N'(t) = 0$. □

Another important result proved in [1], (and queried in [2]), also involves H_1-inversions.

Theorem 1.4 *If T, T′ are any two parallelogram tilings of a semiregular polygon, then it is possible to transform T into T′ by a series of H_1-inversions.* □

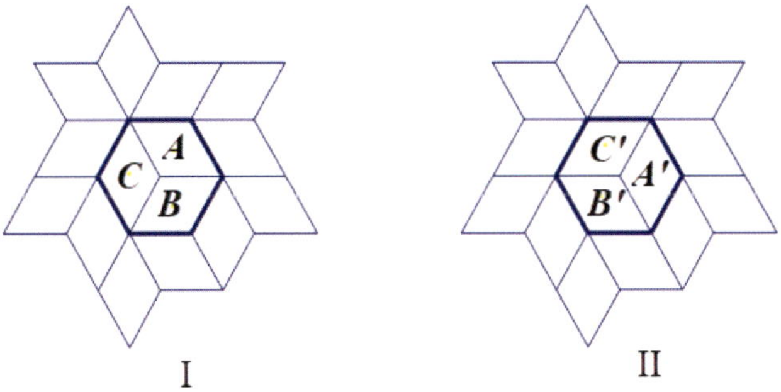

Figure 1.10

In Figure 1.11 Theorem 1.4 is illustrated: the original tiling of H_2 = H(2,2,2) using three P-tiles is transformed by a series of H_1-inversions to the *flower* tiling (1.11(V), given also in Figure 1.1.

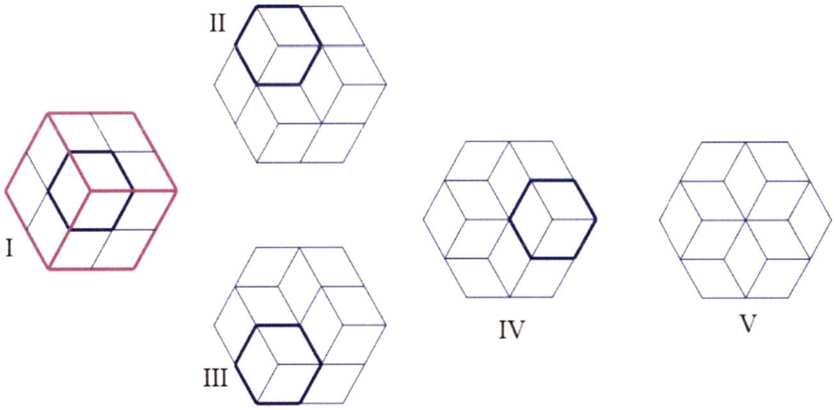

Figure 1.11

Now suppose H = H(p,q,r) is a semiregular hexagon and p,q,r are all even. Then H can be tiled entirely with P-tiles as in Figure 1.11(I). The P-tiles form a nbd dissection of H and thus H, with this tiling, is nilpotent. Since, by Theorem 1.4, any tiling of H by rhombs can be obtained by a series of H_1-inversions, and since by Theorem 1.3 each such H_1-inversion preserves nilpotence, we can conclude that every tiling of H by rhombs is nilpotent. We have proved:

Theorem 1.5 *If H = H(p,q,r) is a semiregular hexagon with p,q,r all even, then every tiling of H by rhombs is nilpotent.* □

Actually Theorems 1.3 and 1.4 yield the following more general result.

Theorem 1.6 *Let P be a semiregular polygon tiled with rhombs. If each side of P has even length then P is nilpotent.* □

1.4 Hexagons with Sides of Mixed Parity

Let P be a semiregular polygon tiled with rhombs. A *stream* in P is a sequence of rhombs with mutually parallel edges. In Figure 1.12(I) the streams of a semiregular polygon are indicated by curves which run through the tiles belonging to each stream.

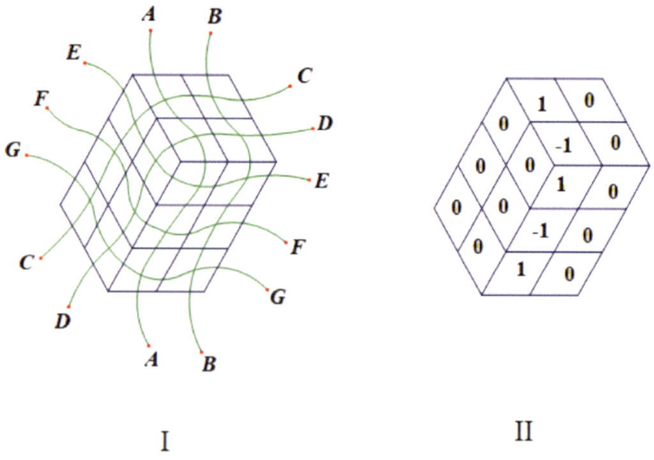

Figure 1.12

Streams A,B,C,D have length 5 and streams E,F,G have length 4. In general, the streams in H = H(p,q,r) have lengths p+q, p+r and q+r. In case the parameters p,q,r are of mixed parity, then H must contain an odd length stream. In Figure 1.12(II) an odd length stream has been selected and assigned integer values alternating between 1 and -1. The remaining tiles are all assigned 0 values. The resulting assignment is locally nilpotent, but $\Sigma H = 1$, so H is not nilpotent. If P is any semiregular polygon tiled with rhombs, and which has an odd length stream, then the same procedure can be used to produce a locally nilpotent assignment for P but such that P itself is not nilpotent. Thus we have:

Theorem 1.7 *Let P be a semiregular polygon tiled with rhombs, and suppose P contains an odd length stream. Then P is not nilpotent.* □

In particular, any hexagon with sides of mixed parity contains an odd length stream and thus is not nilpotent.

1.5 Exercises

1. If P is a semiregular polygon tiled with rhombs, and each side of P has even length, show that the only way to obtain a nbd dissection of P is to use the parallelogram tiles (P-tiles) exclusively.

2. Construct R(60) tilings of H_5 with the following compositions. (I) $H_1^4C^3P^9A^3$; (II) $H_1^6P^9A^3S^1$.

3. Using a series of R(60) inversions convert the tiling of H_5 in Figure 1.7 to the $H_1^1C^{18}$ tiling in Figure 1.8.

4. Use H_1-inversions to transform tiling in Figure 1.13(I) to the one in 1.13(II). Provide nbd dissections for these hexagons and use R(60) inversions to transform 1.13(I) into 1.13(II).

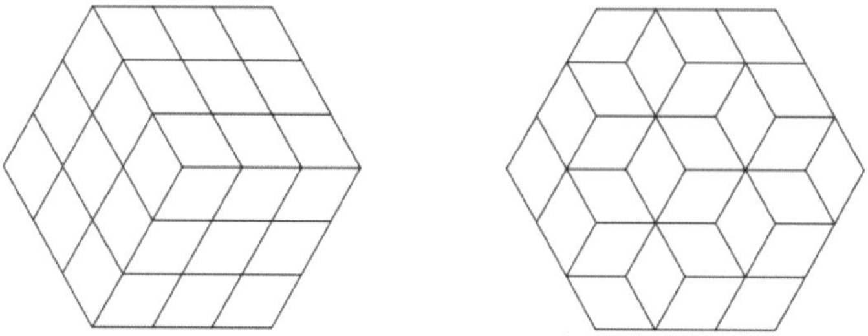

I II

Figure 1.13

5. Prove or disprove Conjecture 1.2.

6. If H = H(p,q,r) is tiled with rhombs. The *order* of H is the number n of rhombs used in the tiling. Show that n = pq + pr + qr.

7. If p,q,r are odd, and H =(H(p,q,r)) is tiled with rhombs, let η(H) denote the number of R(60) tiles used in any nbd dissection of H. Show that η(H) = (1/4)(n+1) where n denotes the order of H.

8. Note as in Figure 1.13, that in the tiling of a hexagon by 60 degree rhombs, the rhombs can occur in one of three orientations. Given a particular semiregular hexagon H, show that in every rhomb tiling of H the numbers of rhombs in each orientation remains constant.

9. Can Theorem 1.7 be strengthened as follows?

Conjecture 1.8 *Let P be any polyrhomb. If P contains a stream with odd length, then P is nonnilpotent.*

CHAPTER 2

Polygons with More than 6 Sides

2.1 General Results

In Theorem 1.7 it was shown that a semiregular polygon P tiled with rhombs must be nonnilpotent in case P contains a stream of odd length. The following is a simple Corollary.

Corollary 2.1 *Let P be a semiregular polygon tiled with rhombs. If P contains a side of even length and a side of odd length, then P is nonnilpotent.*

Proof. Let the sequence S = ($s_1, s_2, \ldots, s_n$) denote the side lengths in a semiperimeter of P. The sum of any n-1 of these values represents a stream length in P. Let E denote the subsequence of S consisting of even values, and let D denote the subsequence of S consisting of odd values. We are assuming that E, D are nonempty. Suppose |D| is odd and let e ∈E. Then the elements in D∪E −{e} form a subsequence of S with n-1 elements and with ∑(D∪E −{e}) = ∑D + ∑E - e an odd value.

On the other hand, if |D| is even, let d∈D. Then the elements D∪E −{d} form a susequence of S with length n-1 and with ∑(D∪E −{d}) = ∑D + ∑E - d an odd value. Thus in all cases P contains a stream of odd length and by Theorem 1.7 is nonnilpotent. □

Another simple Corollary follows.

Corollary 2.2 *Let P be a semiregular polygon tiled with rhombs and suppose* $S = (s_1, s_2, \ldots, s_n)$ *is the sequence of side lengths constituting a semiperimeter of P. If each* s_i *is odd and n is even, then P is nonnilpotent.*

Proof. Since n is even the sequence $(s_1, s_2, \ldots, s_{n-1})$ is of odd length. Since each s_i is odd, the sum $s_1 + s_2 + \ldots + s_{n-1}$ is odd. This sum also represents the length of a stream in P, and thus by Theorem 1.7 P is nonnilpotent. □

If in these Corollaries each s_i is odd and *n is also odd* then a quite different result is obtained. As a preliminary we will establish the following result. The notation $P_1(n)$ will be used to denote a regular polygon with n sides and one edge per side.

Theorem 2.3 *If n/2 is odd then* $P_1(n)$ *has a tiling with a nbd dissection.*

Proof. In Figure 2.1(I, II) nbd dissections are indicated for $P_1(10)$ and $P_1(14)$. We refer to the tilings used as *polar* with the indicated point p as the *pole.* A distinguishing feature of a polar tiling of $P_1(n)$ is the covering of the pole by (n/2)-1 tiles which can be paired to form (1/2)((n/2)-1) H_1 hexagons.

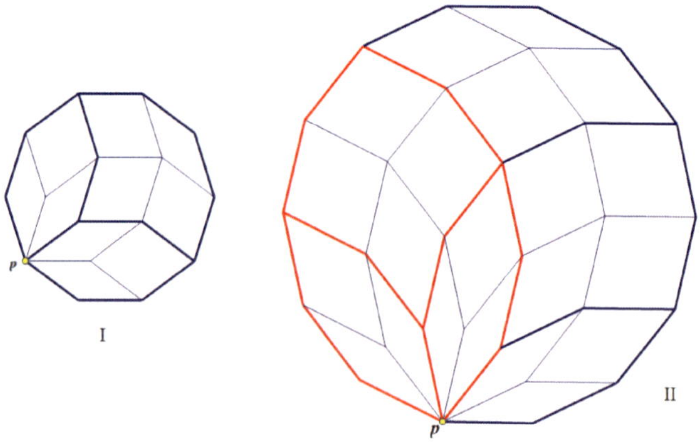

Figure 2.1

In Figure 2.1(II) observe that the polar $P_1(14)$ is an extension of a $P_1(10)$ indicated in red. In general, the polar tiling of $P_1(n)$ consists of a polar tiling of $P_1(n\text{-}4)$ and an extension with a nbd dissection consisting of a single polar H_1 and (1/2)((n/2)-1) - 1 regions which we call *isosquares* since thay are isomorphic as tilings to squares, in this case squares of order 2. Thus an inductive argument can be used to obtain a nbd dissection of the polar tiling of $P_1(n)$. □

As a consequence of Theorem 2.3 we have that the polar tiling of $P_1(n)$ is nilpotent whenever (n/2) is odd. Using Theorems 1.3, 1.4 the following Corollary is immediate.

Corollary 2.4 If n/2 is odd then every tiling of $P_1(n)$ is nilpotent. □

In Figure 2.2 a retiling of the 14-gon in Figure 2.1(II) is illustrated. By Corollary 2.4 it must be nilpotent although it has no nbd dissection.

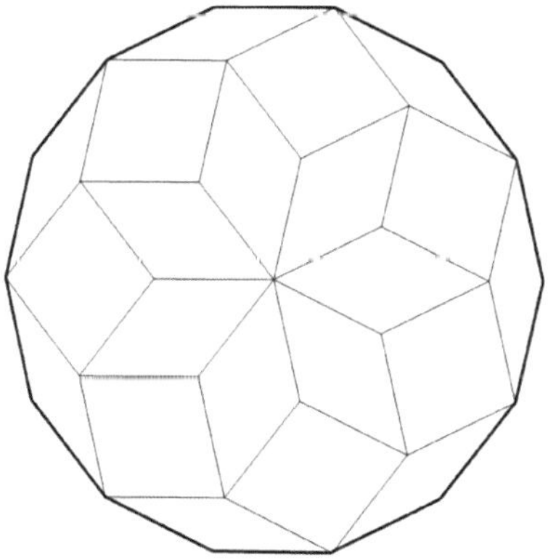

Figure 2.2

A further generalization of Theorem 2.3 is not hard to obtain.

Theorem 2.5 *Let P be a semiregular polygon whose semiperimeter consists of an odd number of sides each with odd length. Then P is nilpotent.*

Proof. Starting with the polar tiling of $P_1(n)$ with n/2 odd, we can extend any opposite pair of sides by two units. When this extension is tiled, it will have a dissection into isosquares. Since the polar tiling of $P_1(n)$ has a nbd dissection, the extended polygon will also have a nbd dissection. It will have the same odd number of sides as $P_1(n)$ and each side will continue to have odd length. Another pair of opposite sides can then be selected and another 2 unit extension constructed. These extensions can be repeated until a tiling of the semiregular polygon P is obtained which has a nbd dissection. It will then follow from Theorems 1.3, 1.4 that P is nilpotent. □

In Figure 2.3 a polar tiling of $P_1(10)$ is extended according to the method of Theorem 2.5, by two units in three directions to obtain the decagon which we denote by D(1,3,3,1,3) indicating the sequence of side lengths. With Theorem 2.5 a full determination of nilpotent semiregular polygons has been achieved. The results are summarized in Theorem 2.6.

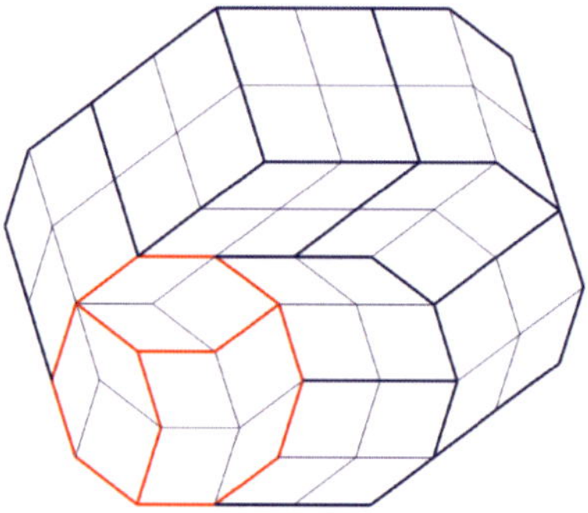

Figure 2.3

Theorem 2.6 *Let P be a semiregular polygon tiled with rhombs and suppose P has n sides. (i) If every side of P has even length then P is nilpotent. (ii) If P has two sides with lengths of opposite parity then P is not nilpotent. (iii) If n is twice an odd number, and each side has odd length, then P is nilpotent. (iv) If n is even and each side of P has odd length, then P is not nilpotent.* □

2.2 Exercises

1. Transform the tetradecagon in Figure2.1(II) to the tetradecagon in Figure 2.2 using H_1 - inversions.

2. Figure 2.4 features two nbd dissections of the decagon $P_1(10)$. The rhombs involved have interior angles which are multiples of 36°. Thus the nbds belong to R(36). Use R(36) inversions to transform the dissection in A to the dissection in B.

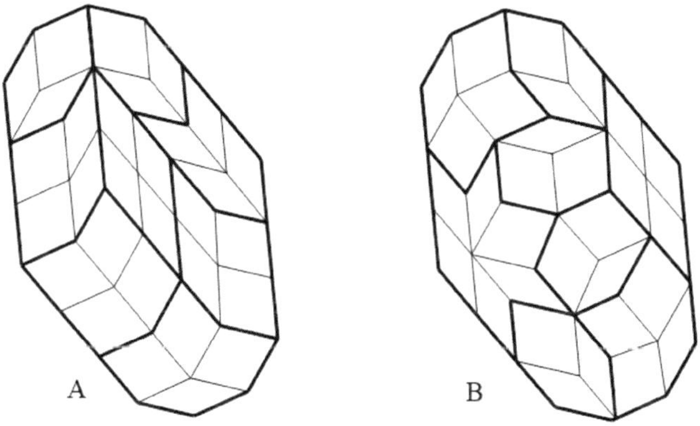

Figure 2.4

Conjecture 2.7: Let D = D(p,q,r,s,t) be a semiregular decagon with p,q,r,s,t all odd and let T_1, T_2 be two tilings of D with rhombs such that T_1 has a nbd dissection N_1, and T_2 has a nbd dissection N_2. Then there exists a sequence of R(36) inversions which transform N_1 into N_2.

3. Prove or disprove Conjecture 2.7.

4. Let D = D(p,q,r,s,t) be a semiregular decagon with p,q,r,s,t all odd. If a tiling T of D by rhombs has a nbd dissection N, let $\eta(N)$ denote the number of nbd's in N.

Conjecture 2.8: *For every nbd dissection N of D, $\eta(N) = (1/4)(n+2)$ where $n = pq+pr+ps+pt+qr+qs+qt+rs+rt+st$ denotes the order of D, i.e., the number of rhombs used in any tiling of D by rhombs.*

Prove or disprove Conjecture 2.8. Note that this will follow from the Conjecture 2.7 in (4) since R(36) inversions preserve the number of nbd's.

5. Two nbd dissections of the regular decagon D(3,3,3,3,3) are illustrated in Figure 2.5. Use R(36) inversions to transform 2.5(I) into 2.5(II).

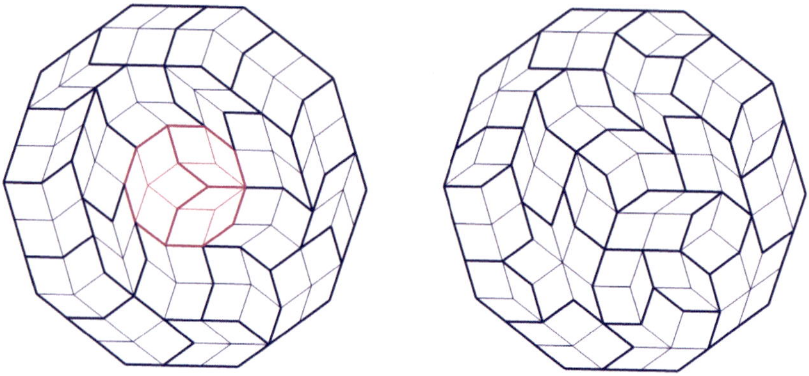

I

II

Figure 2.5

CHAPTER 3

Nilpotent Plane Graphs

3.1 Semiregular Polygons and Plane graphs

Let P be a semiregular polygon tiled with rhombs and let $\mathbf{G}_P$ denote the graph formed by the vertices and sides of the tiles. The *dual* graph $\mathbf{G}_P^d$ is defined as follows. If R= $\{R_1, \dots, R_n\}$ is the set of rhombs used to tile P, let $\{p_1, \dots, p_n\}$ denote the vertex set of G_P^d where for each i, p_i corresponds to R_i. In case tiles R_i, R_j have an edge in common, include p_ip_j as an edge of G_P^d. The graph $\mathbf{G}_P^d$ which corresponds to the hexagon in Figure 1.2 is illustrated in Figure 3.1.

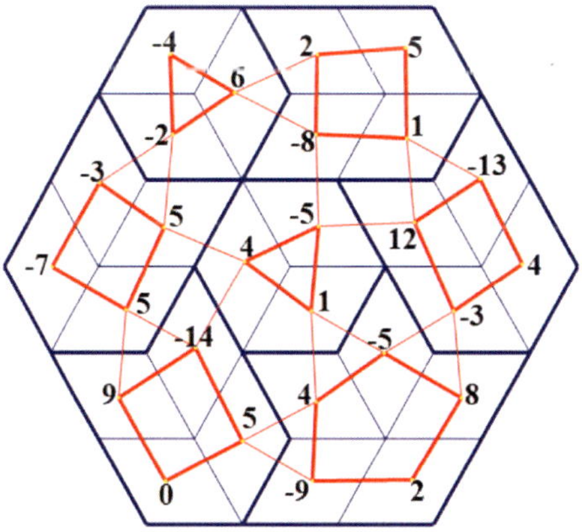

Figure 3.1

The finite regions in $\mathbf{G}_P^d$ correspond to nbds in P. The regions of $\mathbf{G}_P^d$ which correspond to the nbd dissection of P are indicated in Figure 3.1 in bold. A region with m sides in $\mathbf{G}_P^d$ corresponds to a nbd in P consisting of m rhombs. Every interior vertex of $\mathbf{G}_P^d$ has degree 4 since every interior tile of P shares each of its edges with exactly 4 rhombs. Since no more than 6 and no less than three 60° rhombs can surround a vertex of P, the regions in $\mathbf{G}_P^d$ can only have 3, 4, 5 or 6 sides.

3.2 An Interesting Example

A plane graph W has a *locally nilpotent assignment from an abelian group G* in case elements from G are assigned to the vertices of W so that the values assigned to vertices which surround each finite region of W have sum 0. If for every such assignment the sum of the values assigned to the entire vertex set of W is 0, then we say W is *nilpotent*. In case P is a semiregular polygon tiled with rhombs, then P is nilpotent iff the graph $\mathbf{G}_P^d$ is nilpotent. Consider the example given in Figure 3.2.

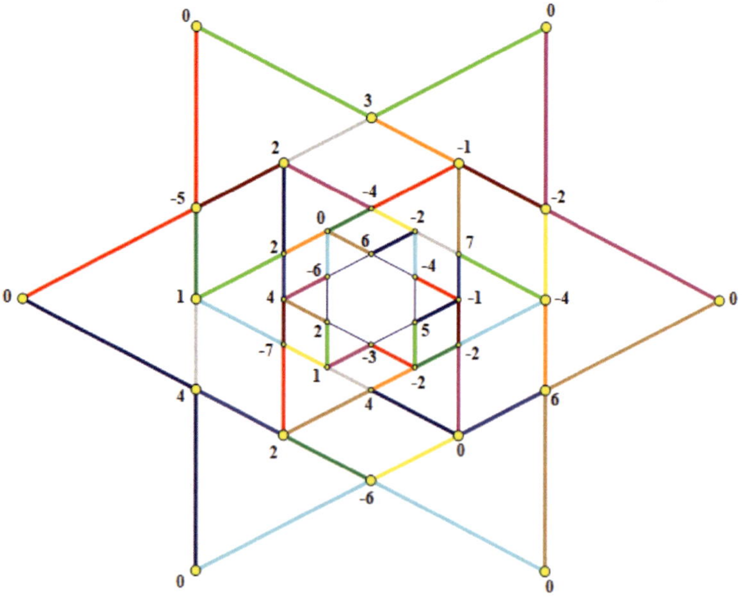

Figure 3.2

In Figure 3.2 edges whose endpoints must have the same sum *for any assignment* are given the same color. The far edges of two adjacent quadrilaterals must have the same color; the far edges of two triangles with a point in common must have the same color; in case two quadrilaterals share edges with two touching triangles, then the far edges of the quadrilaterals must have the same sum. Each of these situations can be found in the above plane graph. In plane graphs constructed with mostly triangular and quadrilateral regions, sequences of edges and points which have the same value are useful in their analysis. We shall refer to such sequences as *edge-point trails.* Various properties of this graph will be proved with reference to Figure 3.3.

(0) Each subgraph bound by a star is nilpotent.

(1) Each subgraph bound by a hexagon is nilpotent.

(2) The subgraphs bound by triangles (J_1, J_3, J_5), (J_2, J_4, J_6), (L_1, L_3, L_5), (L_2, L_4, L_6), (X_1, X_3, X_5), (X_2, X_4, X_6) are nilpotent.

(3) The points $\{X_1, X_2, X_3, X_4, X_5, X_6, M_1, M_2, M_3, M_4, M_5, M_6\}$ which bound the infinite region have sum 0.

(4) $X_1 = X_2 = X_3 = X_4 = X_5 = X_6 = 0$.

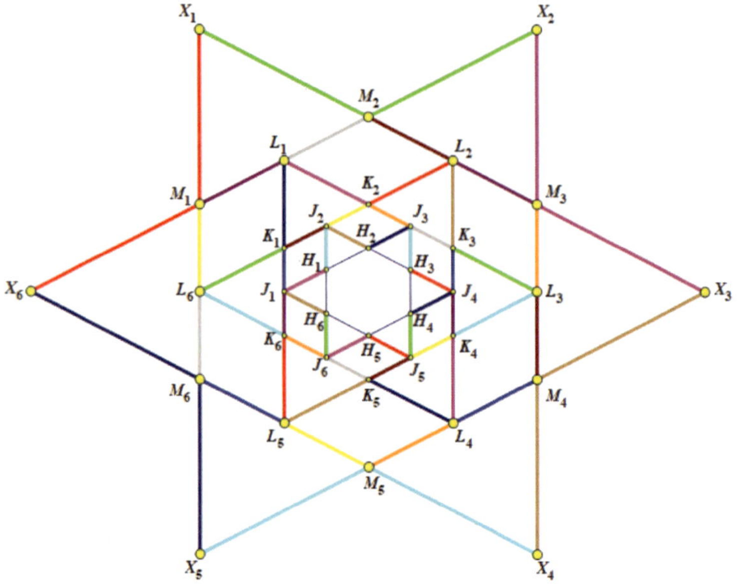

Figure 3.3

First note that the subgraph induced by the J and H points corresponds to the nilpotent hexagon in Figure 1.1 and thus is itself nilpotent. The subgraph with triangular boundary (J_1, J_3, J_5) has a vertex cover consisting of three small disjoint triangles, as does the triangular region with boundary (J_2, J_4, J_6), so both of these subgraphs are nilpotent. (Note that a semiregular polygon P tiled with rhombs has a nbd dissection iff the graph $\mathbf{G}_P{}^d$ has a *2-factor*, i.e., a vertex cover consisting of disjoint cycles. A 2-factor for the graph in Figure 3.1 is indicated in bold. A plane graph W with a 2-factor each of whose cycles is a region of W is necessarily nilpotent.)

The point sets {J_1, J_3, J_5, J_2, J_4, J_6} and {H_1, H_2, H_3, H_4, H_5, H_6} are *0-sum sets* since {H_1, H_2, H_3, H_4, H_5, H_6} are the boundary points of a region, and these together with {J_1, J_3, J_5, J_2, J_4, J_6} form the above mentioned nilpotent hexagon. To show that the hexagonal subgraph with boundary formed by the K and J points is nilpotent, consider the following equations.

(1) $K_1 + J_1 + J_2 + H_1 = 0$
(2) $K_2 + J_2 + J_3 + H_2 = 0$
(3) $K_3 + J_3 + J_4 + H_3 = 0$
(4) $K_4 + J_4 + J_5 + H_4 = 0$
(5) $K_5 + J_5 + J_6 + H_5 = 0$.

Summing these equations gives

(6) $(K_1 + \ldots +K_6) + 2(J_1 + \ldots + J_6) + (H_1 + \ldots + H_6) = 0$, and thus
(7) $(K_1 + \ldots +K_6) + (J_1 + \ldots + J_6) + (H_1 + \ldots + H_6) = 0$.

Note that

(8) $(K_1 + \ldots +K_6) = 0$ follows from (7).

It follows that the hexagonal subgraph with corners $K_1, \ldots ,K_6$ is nilpotent since it is composed of the 0-sum set $\{K_1, \ldots ,K_6\}$ and the nilpotent star consisting of the J and H points.

Now in the triangle with boundary $\{L_1, L_3, L_5\}$ the three quadrilateral regions formed by an L, J and two K points yield the equations

(9) $L_1 + K_1 + J_2 + K_2 = 0$
(10) $L_3 + K_3 + J_4 + K_4 = 0$
(11) $L_5 + K_5 + J_6 + K_6 = 0$.

Summing these equations yields

(12) $(L_1 + L_3 + L_5) + (K_1 + \ldots + K_6) + (J_2 +J_4 + J_6) = 0$.

Since the triangular subgraph with corners $\{J_2,J_4, J_6\}$ is nilpotent and $(H_1 + \ldots + H_6) = 0$, we have

(13) $J_2 +J_4 + J_6 = 0$, and similarly
(14) $J_1 + J_3 + J_5 = 0$.

Then from (12) we obtain

(15) $L_1 + L_3 + L_5 = 0$ and similarly
(16) $L_2 + L_4 + L_6 = 0$.

It now follows that the star with points $\{L_1, \ldots, L_6\}$ is nilpotent. The 0-sum equations obtained from the six M, K and L quadrilaterals yield similarly that $M_1 + \ldots + M_6 = 0$. It then follows that the subgraph with hexagonal boundary determined by the points M_1, … , M_6 is nilpotent.

The 0-sum quadrilaterals which lie in the large triangle with vertices X_1, X_3, X_5 yield the equations

(17) $X_1 + M_1 + M_2 + L_1 = 0$
(18) $X_3 + M_3 + M_4 + L_3 = 0$
(19) $X_5 + M_5 + M_6 + L_5 =0$.

When these are summed we obtain

(20) $(X_1 + X_3 + X_5) + (M_1 + \ldots +M_6) + (L_1 + L_3 + L_5) = 0$ which results in
(21) $X_1 + X_3 + X_5 = 0$, and similarly
(22) $X_2 + X_4 + X_6 = 0$.

It now follows that the entire subgraphs bound by the triangles (X_1, X_3, X_5) and (X_2, X_4, X_6) are nilpotent and in fact the entire graph in Figure 3.3 is nilpotent.

Since the boundary edges incident with each of the M points have the same color, the sum of each of these pairs of edges must be the same. Thus, for example $M_1 + X_1 = M_1 + X_6$ with the consequence that $X_1 = X_6$. Similarly $X_6 = X_5$ and ultimately

(23) $X_1 = X_2 = X_3 = X_4 = X_5 = X_6$ is obtained.

Then (21), (22) and (23) imply

(1) $X_1 = X_2 = X_3 = X_4 = X_5 = X_6 = 0$.

In general a set of points which must have the same sum k is called a *k-set.* Thus for the graph in Figure 3.3, the set $\{X_1, X_2, X_3, X_4, X_5, X_6\}$ is a 0-set.

3.3 Nilpotent Plane Graphs with Rotational Symmetry

Nilpotent plane graphs like the one in Figure 3.3, which also have rotational symmetry have some common properties which we consider here. Our constructions involve repeated extensions of *0-sum boundaries.* The 0-sum boundaries in Figure 3.3 are the boundaries formed by the (i) H points; (ii) H and J points; (iii) J and K points; (iv) K and L points; (v) L and M points; (vi) M and X points. An extension which takes one 0-sum boundary to a larger one can be of several types including the following.

Suppose B is a 0-sum boundary. (1) attach a triangle to each edge of B. The new points used to create these triangles can be replaced by any path; (2) If $B = (J_1H_1J_2H_2 \dots J_mH_m)$ and $H_1 + \dots + H_m = 0$ and $J_1 + \dots + J_m = 0$ then the addition of new points $K_1, \dots, K_m$ and edges $\{K_iJ_i, K_iJ_{i+1}\} \cup \{K_mJ_m, K_mJ_1\}$ creates the 0-sum boundary $(K_1J_2K_2J_3K_3J_4K_4 \dots J_mK_m)$. In this second type of extension, each point K_i can be replaced by any path; (3) If B can be represented in the form $B = (C_0, C_1, \rho_1, C_2, C_3, \rho_2, C_4, C_5, \rho_3, \dots, \rho_k)$ where the pairs $\{C_i, C_{i+1}\}$ are consecutive points of B, and the ρ_i represent intervening segments of B, then the addition of points $K_1, \dots, K_k$ to form regions $(K_i, C_i, \rho_i, C_{i+1})$ results in a new 0-sum boundary. Here also any path can be substituted for the points K_i.

In Figure 3.3, the extension from the H hexagon to the H and J star is of type (1). The further extension to the J and K hexagon is of type (2). In Figures 3.4(I) and 3.4(II) two more examples of nilpotent graphs featuring 0-sum boundary extensions and rotational symmetry are illustrated.

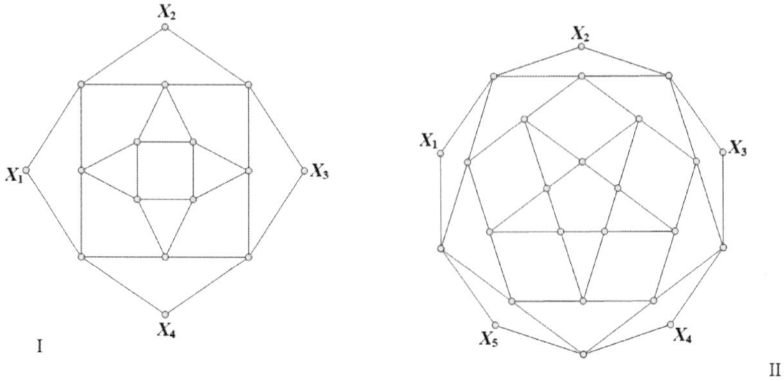

Figure 3.4

The square in the middle of 3.4(I) is the *central polygon* for the construction, and the central polygon for 3.4(II) is a pentagon. We leave it to the reader to show that both graphs are nilpotent and the X-points in each graph are 0-sets.

In Figure 3.5(I,II) a nilpotent graph with 90° rotational symmetry and central square is illustrated. In 3.5(I) a locally nilpotent assignment is made from the group Z of integers. It is important to note that this assignment starts freely with the central square and with its four adjacent squares, but after these the remaining assignments are all forced. Attempts to give a free assignment to a noncentral region and then extend to a locally nilpotent assignment are likely to encounter difficulties.

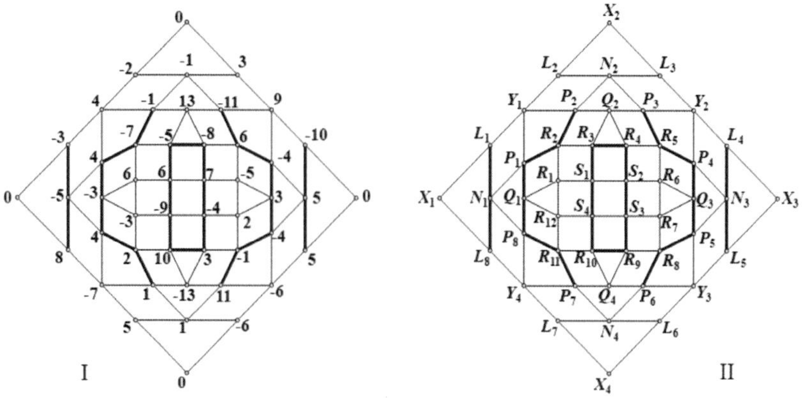

Figure 3.5

We will just prove that the four X-points in 3.5(II) form a 0-set as suggested by the assignment in 3.5(I). The proof of nilpotence using a series of 0-sum boundaries we leave to the reader. We will show that the point sets $\{L_4, N_3, L_5\}$, $\{P_3, R_5, P_4, Q_3, P_5, R_8, P_6\}$ and $\{R_3, R_4, S_1, S_2, S_3, S_4, R_9, R_{10}\}$ indicated in bold all have the same sum. Consider:

(1) $L_4 + N_3 + L_5 = (-Y_4 - P_4 - N_3) + (-P_4 - Q_3 - P_5) + (-N_3 - P_5 - Y_3) =$
$(P_3 + R_5) - N_3 + (R_5 + R_6) - P_5 - N_3 + (R_8 + P_6)$.

Note that $-N_3 + R_5 + R_6 = (P_4 + Q_3 + P_5) + (-P_4 - Q_3) = P_5$, and so from (1) we obtain

(2) $L_4 + N_3 + L_5 = P_3 + R_5 - N_3 + (R_8 + P_6) = P_3 + R_5 + P_4 + Q_3 + P_5 + R_8 + P_6$,

as desired. Now continuing from the right hand side of (2) we have

(3) $(P_3 + R_5) + P_4 + Q_3 + P_5 + R_8 + P_6 = R_3 + (R_4 + S_2) + P_5 + R_{10} =$
$R_3 + (R_4 + S_2) + (-Q_3 - R_7 - R_8) + R_{10} = R_3 + (R_4 + S_2) + R_6 - R_8 + R_{10} =$
$R_3 + (R_4 + S_2) + (-S_2 - S_3 - R_7) + (S_3 + R_7 + R_9) + R_{10} = R_3 + R_4 + R_9 + R_{10} =$
$S_4 + S_3 + S_1 + S_2 = 0$.

We have shown that $L_4 + N_3 + L_5 = 0$, and it follows immediately that $X_3 = 0$. The remaining X-points are then 0 by symmetry. Our main interest will be in constructions which obtain 0-sets in the outer boundary.

If W is a nilpotent plane graph and a point P is joined to W via two edges, the resulting graph is a *point extension* of W. The nilpotent graphs in Figures 3.4, 3.2 and 3.5 were constructed via a series of point extensions from a central pentagon, hexagon and square respectively. This kind of construction insures the existence of a locally nilpotent assignment for W which extends a free assignment for the central polygon. In the above mentioned Figures, all point extensions create either triangular or quadrilateral regions. In Figures 3.6(I,II,III) various nilpotent graphs are constructed with central triangle and rotational automorphism of order 3. The X-points indicate 0-sets.

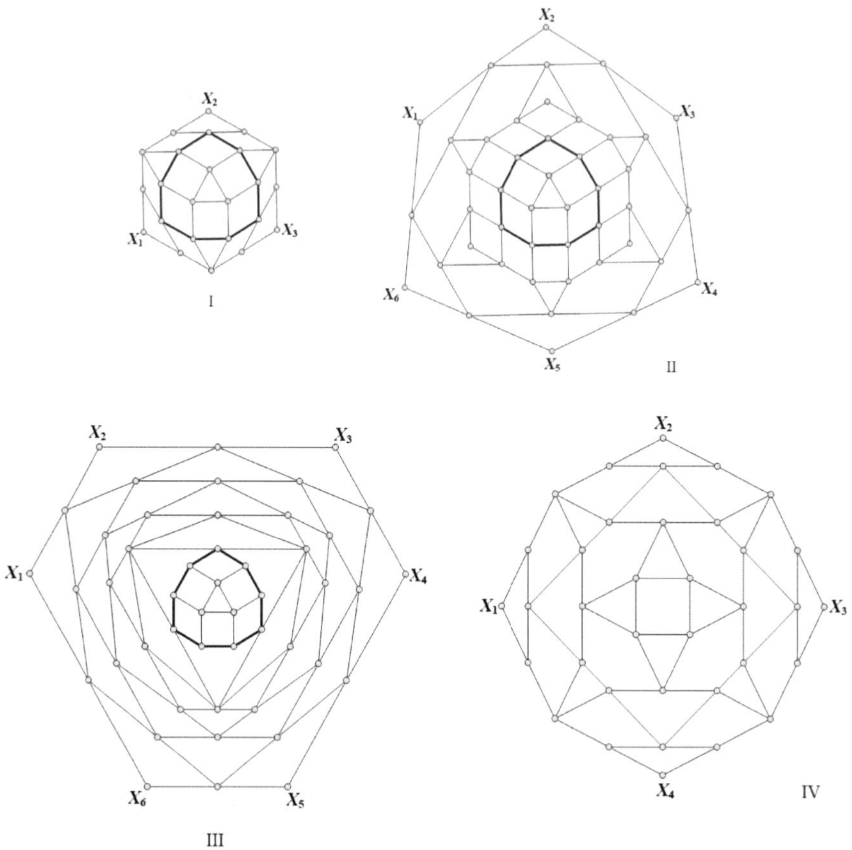

Figure 3.6

In 3.6(I,II,III) point extensions of the indicated central structure are repeatedly employed. In Figure 3.6(IV) the four pentagons represent *2-point extensions* of the central structure.

3.4 Handles

It is also possible to extend a nilpotent graph W by inserting edges which join two boundary points of W. We call such edges *handles*. The first thing to note is that points in a 0-set can be replaced by handles. Thus, for example, the graph in Figure 3.7(I) is obtained from 3.6(I) by replacing each of the 0-points, (represented by X_1, X_2, X_3) by a handle.

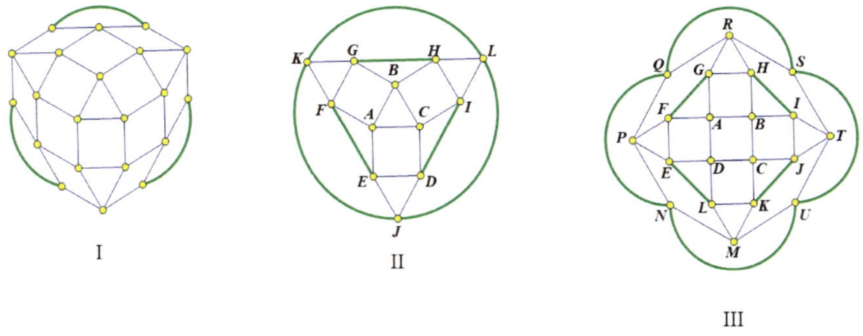

Figure 3.7

All three graphs in Figure 3.7 are 4-regular. By a Theorem of Petersen (see [3]), every 4-regular graph has a 2-factor and if the cycles which constitute the 2-factor are all regions in a plane graph, then the graph is nilpotent. The locally nilpotent assignment for graphs with handles does however provide something to prove. Certainly such an assignment can be trivially achieved by assigning 0 to each point, but we seek the *most general* locally nilpotent assignment. The *degree of freedom* for a plane graph W is the maximum number of vertices which can freely be assigned integers and still be extended to a locally nilpotent assignment for the entire graph W.

The degree of freedom for 3.7(II) is 3: the vertices A,B,G can be assigned any integer values, then the remaining assignments are all forced. For example we can define

(1) C = -A - B
(2) F = -A - B - G
(3) E = -F - A
(4) D = -E - A - C
(5) H = -G - B
(6) I = -H - B - C

With these definitions it is necessary to prove C + D + I = 0. Consider:

C+D+I = (-A - E) + (-H - B - C) = -A + (F+A) + (G+B) - B - C = (F + G)- C = -A -B - C = 0.

The outer handles for 3.7(II) can easily be proved by the reader. We also leave to the reader the analysis of 3.7(III) which has degree of freedom 4: the integer values for A,B,D,F can be chosen freely. Define I, J in terms of the other central values and then prove that B+C+I+J = 0. In Figures 3.8(I,II) two more examples of nilpotent graphs with handles are illustrated.

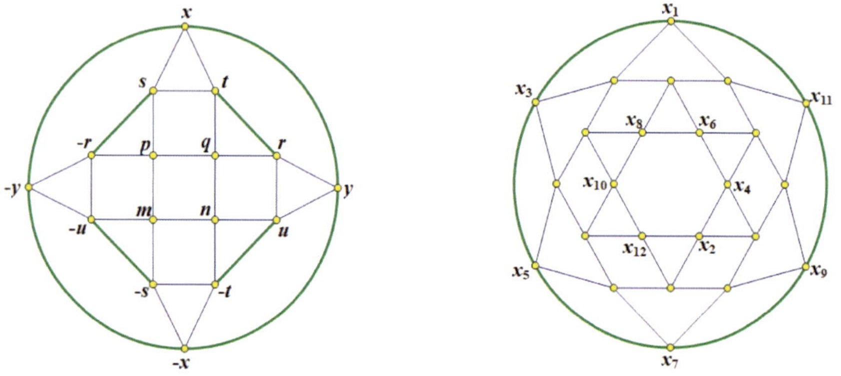

I II

Figure 3.8

In 3.8(I) the points labeled x, -x are two ends of an edge-point trail and thus must have opposite values. Similarly the pairs of points {y, -y}, {s,-s}, {t,-t}, {r,-r}, {u,-u} are ends of an edge-point trail and must have the indicated opposite values. Now considering various 0-sum regions required by a locally nilpotent assignment, we have

(1) x+s+t = 0
(2) y+r+u = 0
(3) x+s - y - r = 0

From (1),(2),(3) u=t is obtained, and then

(4) -x - t + u + y = 0 yields x = y.
(5) x+s - y - r = 0.

Now (1), (5) yield

(6) y+r+t = 0, and then (2),(6) yield
(7) u = t, an then by symmetry we obtain
(8) s = -u.

From the lower left outer quadrilateral we have

(9) -y - u - s - x = 0, and combining with (8) we obtain
(10) x+y = 0, which along with (4) yields
(11) x = y = 0.

Considering the outer quadrilaterals we obtain r = s, and then the inner triangles force m = n = p = q = 0. The end result is that 3.8(I) has degree of freedom equal to 1, with only one free choice for the value of s.

In 3.8(II) the points labeled x_1, … , x_{12} represent, in sequence, the members of equal value in an edge-point trail. The central hexagon yields $x_2 + x_4 + x_6 + x_8 + x_{10} + x_{12} = 0$, and thus $x_2 = x_4 = x_6 = x_8 = x_{10} = x_{12} = 0$. The remaining vertices must then also be 0 in order to achieve a locally nilpotent assignment. We say that 3.8(II) is *trivially nilpotent*, (with degree 0), since local nilpotence can only be obtained by assigning 0 to every point. The graphs 3.4(I), 3.4(II), 3.8(II) have similar structure: a central regular polygon surrounded by triangles, then by quadrilaterals and then again by quadrilaterals and finally by triangles. In case the central polygon has 6,7,8 or 9 sides a trivially nilpotent graph is obtained. When the central polygon is a decagon as in Figure 3.9, an interesting nontrivial nilpotent graph is obtained.

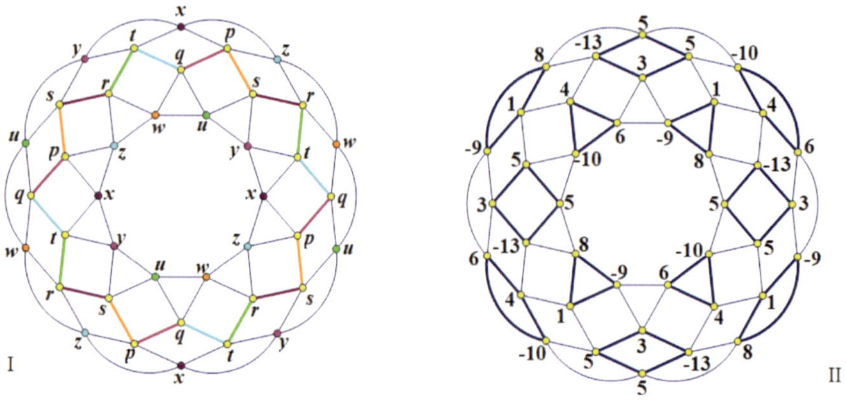

Figure 3.9

In 3.9(I) edge-point trails are used to determine points which must be assigned identical values. For this graph a total of 10 different values is possible, with each value occurring exactly 4 times. Once x, y, z, w have been chosen freely, the value of u must be -x -y -z -w in order for the central decagon to have 0-sum. The values for the remaining points are all forced, and thus 3.9(I) has degree of freedom 4. In 3.9(II) a particular locally nilpotent assignment is illustrated which follows the pattern indicated in 3.9(I). The bold edges form a 2-factor consisting of regions, and thus the entire graph is quickly seen to be nilpotent.

3.5 Exercises

1. Show that the graphs in Figure 3.4(I,II) are nilpotent and the X-points in each graph are 0-sets.

2. Show that the graph in Figure 3.5(II) is nilpotent.

3. Show that the graphs in Figure 3.6 are nilpotent and the X-points in each graph are 0-sets.

4. Show that the graph in Figure 3.7(III) has degree of freedom 4 and prove that a locally nilpotent assignment is achieved.

5. Let C be a regular central polygon with n sides which is extended first by triangles, then by quadrilaterals, then again by quadrilaterals, and finally by triangles, as in the examples in Figures 3.8(II), 3.9. For which values of n does the resulting graph have a nontrivial locally nilpotent assignment from the group Z of integers?

6. Consider the sequence of graphs: 3.7(II), 3.8(I) followed by 3.10(I,II) illustrated below. Show that these graphs are nilpotent and each eventually attain a 0-sum boundary whose alternate values form a 0-set. If the sequence is continued, is a 0-set always attained?

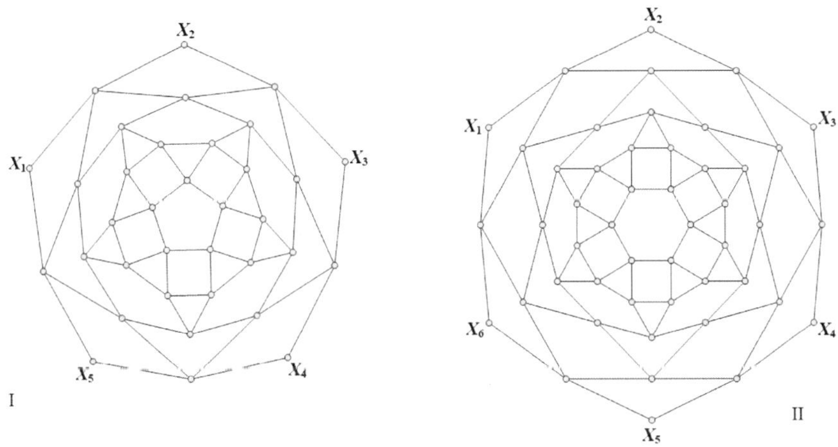

Figure 3.10

7. Let W be a plane graph with a locally nilpotent assignment from the set Z of integers, and suppose W is regular of degree 4. By [3] W has a 2-factor. Can a 2-factor always be found each of whose constituent cycles has 0-sum?

CHAPTER 4

Free Generating Sets in G_p and $G_p^{\ d}$

4.1 Introduction

Let W=(V,E) be a plane graph with vertex set V and edge set E, and suppose it is desired to construct a locally nilpotent assignment from Z to the vertex set V. If R is a finite region of W bound by the cycle of vertices (p_1, p_2, ... , p_n) and if any n-1 of these vertices have been assigned integer values then the value of the remaining vertex of the cycle is forced in order to obtain 0-sum. If say vertices p_1, p_2, ... , p_{n-1} have been assigned integer values, then {p_1, p_2, ... , p_{n-1}} *generates* p_n. A *free generating set* in W is a subset of V whose integer values can be freely assigned, and when such assignments are made, the integer values for the remaining vertices of W are all uniquely determined.

In Chapter 3, free generating sets were determined for the plane graphs in Figures 3.7, 3.8 and 3.9. There are a number of interesting problems concerning free generating sets in plane graphs, regardless of whether or not the graph is nilpotent: (1) constructing a free generating set; (2) determining all free generating sets; (3) do all free generating sets for a given plane graph, have the same cardinality? (4) does every member of a given class of plane graphs have a free generating set with the same cardinality?

4.2 Free Generating Sets for Graphs Determined by Tilings with rhombs

Let P be a semiregular polygon tiled with rhombs, and let $\mathbf{G}_P$ denote the corresponding graph and $\mathbf{G}_P{}^d$ the dual whose vertices correspond to tiles of P and which contains an edge xy if the tiles corresponding to x, y have a common edge in P. In this Section we study the classes of graphs {$\mathbf{G}_P$: P is a semiregular polygon with semiperimeter s} and {$\mathbf{G}_P{}^d$: P is a semiregular polygon with semiperimeter s}. If x, y are two vertices of G_P, the *distance* from x to y is the length of a shortest path from x to y. An x, y path which achieves the shortest distance is a *distance path.* A *diameter* of G_P is a distance path which joins two opposite corner points of P. In [1] it is shown that every diameter of G_P contains exactly s edges where s denotes the semiperimeter of P, (which equals the number of streams in P).

In Figure 4.1 points x, y are opposite corners of the semiregular octagon P which has semiperimeter 12. A typical diameter with s = 12 edges joining x and y is indicated in bold. We will show that the s+1 points which belong to the diameter form a free generating set for G_P. In Figure 4.1 the order of generation is indicated by the vertex labels: points in the generating set are labeled 0; points labeled 1 are the ones immediately generated by 0-points; points labeled 2 are ones which are generated by points labeled 0, 1, etc.

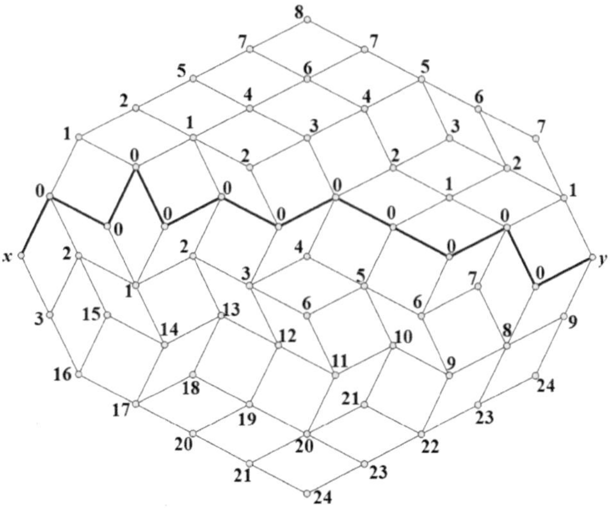

Figure 4.1

Theorem 4.1 *Let P be a semiregular polygon tiled with rhombs and let ρ denote any diameter of G_P. Then the vertices of ρ form a free generating set for G_P.*

Proof. Let R_1 denote the set of tiles on one side of ρ which share an edge of ρ. These are necessarily fewer in number than the number (s) of edges which constitute ρ. As a consequence there must exist at least one tile in R_1 which shares two edges with ρ and thus its fourth point is generated by three points of ρ. Label the points thus immediately generated by ρ with 1 as in Figure 4.1. If we just consider the top side of the x,y diameter ρ in Figure 4.1, we find that R_1 has 7 tiles. Since this is 4 less than the semiperimeter, there will be 4 points to label 1.

Now form a new path ρ_1 which joins points labeled 1 and uncovered points of ρ, (labeled 0). Let R_2 denote the set of tiles *on top* of ρ_1 which share an edge with ρ_1. These will be fewer in number than the number of edges in ρ_1, and thus there must exist at least one tile in R_2 which has two edges in common with ρ_1.

There will be fewer tiles *on top* of ρ_1 which share an edge with ρ_1, than there are edges in ρ_1. As a consequence, the fourth point, (the one not on ρ_1) will join the set of generated points. The collection of points generated in this manner by points labeled 0, 1 are indicated in 4.1 with label 2. The argument now repeats starting with the path ρ_2 consisting of points labeled 0, 1, 2. Ultimately, every point on the top side of ρ is generated, and similarly every point *below* ρ is generated. □

The path which joins opposite corners and uses only boundary edges as in Figure 4.2, is also a diameter. Its vertices form a free generating set with the order of generation indicated in the Figure. Although the same polygon with the same tiling occurs in 4.1 and 4.2, the generating set in 4.1 is more efficient in terms of the number of iterations required: 24 in 4.1 and 31 in 4.2.

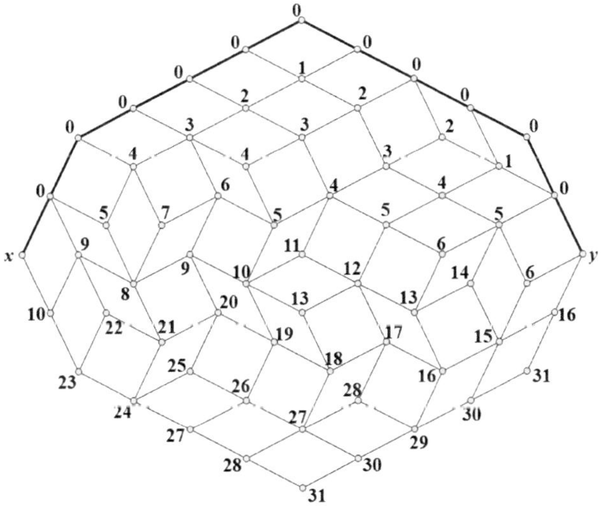

Figure 4.2

Of course three integer values assigned to three corners of a rhomb will force the value for the fourth corner point, (to obtain a 0-sum). Shifting one of the assigned values to the fourth unassigned point will have the same effect of achieving a 0-sum. This type of shift can be applied repeatedly to a diametric free generating set with the result that a new free generating set is obtained. Examples of this construction are illustrated in Figure 4.3.

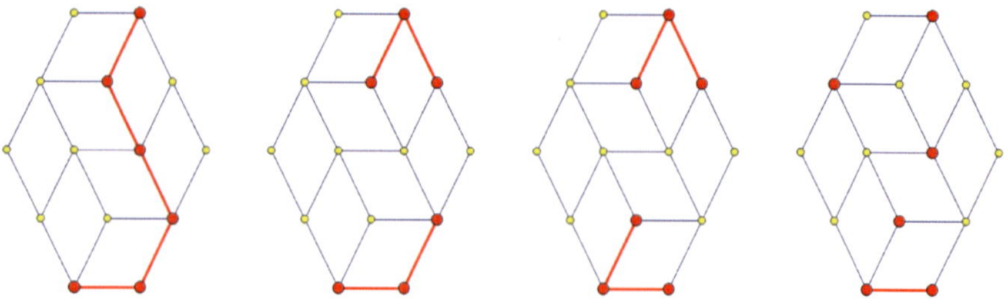

I II III IV

Figure 4.3

A diametric free generating set is indicated in Figure 4.3(I). Figure 4.3(II) is obtained from 4.3(I) by a single shift, and 4.3(II) is obtained by a further shift. 4.3(IV) represents a slightly different construction: single points belonging to the original diameter are shifted to opposite points in the same tile stream. In these examples all free generating sets have the same number of points. This is to be expected as we prove next.

Theorem 4.2 *Let P be a semiregular polygon with semiperimeter s. Then for any tiling of P by rhombs, every free generating set for G_P has exactly s+1 elements.*

Proof. Let e denote the number of edges in G_P and let r denote the number of rhombs used to tile P. That the parameters e, r are independent of the tiling we leave as an exercise. Suppose the semiperimeter consists of n sides with lengths $\{s_1, s_2, \ldots, s_n\}$. Then $s = s_1 + \ldots + s_n$ and $r = (\Sigma s_i s_j : i,j \in \{1, ,n\}, i \neq j)$. For each i there are s_i streams each containing $s_i(s+1-s_i)$ edges. The edge sets comprising the streams form a partition of the set of all edges in G_P and thus

(1) $e = \Sigma\{s_i(s+1-s_i) : i \in \{1, .., n\}\}$.

Consequently

(2) $e - 2r = \Sigma\{s_i(s+1-s_i) : i \in \{1, .., n\}\} - 2(\Sigma s_i s_j : i,j \in \{1, ,n\}, i \neq j)$
(3) $e - 2r = s^2 + s - \Sigma\{s_i^{\,2} : i = 1, \ldots, n\} - 2(\Sigma s_i s_j : i,j \in \{1, ,n\}, i \neq j)$
(4) $e - 2r = s^2 + s - (s_1 + s_2 + \ldots + s_n)^2 = s$.

If v denotes the number of vertices in G_P then Euler's formula [4] yields v = e-r+1. Let x denote the number of elements in a free generating set for G_P, so v-x denotes the number of remaining vertices in G_P. Represent these v-x vertices by variables and form r 0-sum equations, one for each region of G_P. These equations are linearly independent, and equating the number of equations with the number of variables yields v-x = r or x = v-r = e-2r+1 = s+1. Thus we would expect a free generating set to have s+1 elements. □

To determine free generating sets for $\mathbf{G}_P^d$ the following Lemma will be useful.

Lemma 4.3 (two streams Lemma) *Let P be a semiregular polygon tiled with rhombs and let s denote the semiperimeter of P. Let x. y be opposite corner vertices of* $\mathbf{G}_P$ *and let A, B denote tiles incident with x, y respectively. Let* S_A, S_B *denote intersecting tile streams containing A, B respectively, and let W denote the unique tile common to both streams. The tile path ρ which follows* S_A *from A to W and then follows* S_B *from W to B has exactly s-1 tiles.*

Proof. (See Figure 4.5)

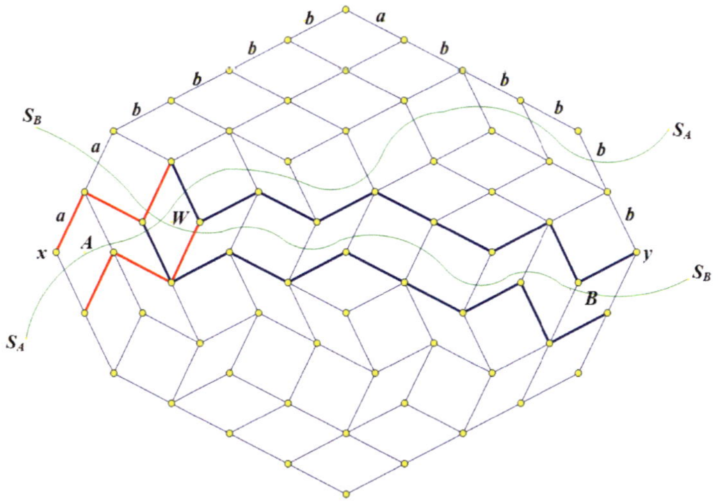

Figure 4.5

The path σ along the edge of ρ which joins x, y is a diameter of $\mathbf{G}_P$ and thus has length s. Each of the s streams of P will cross σ in exactly one edge. If the section of ρ from A to W, (including tile W) consists of m tiles, then these tiles will belong to m distinct streams of P. Similarly, if the section of ρ from W to B contains n tiles, then these tiles will belong to n distinct streams of P. Since σ has s edges we have m+n = s. In Figure 4.5, the streams through the first section of ρ are indicated with an "a" on their initial edge, and the streams through the remainder of σ are indicated with a "b" on their initial edge. Only the tile W where S_A, S_B intersect will belong to both an "a" and a "b" stream. The number of tiles in ρ will then be m+n - 1 = s-1. □

By Lemma 4.3 the existence of a tile path of length s-1 which joins two opposite corner tiles of P is demonstrated. However, a free generating tile set need not have the exact structure given in Lemma 4.3. In Figure 4.6(I) a single tile stream serves as a free generating set, and in Figure 4.6(II) a free generating path consisting of sections of four tile streams is illustrated. In both cases the generating path has exactly s - 1 tiles.

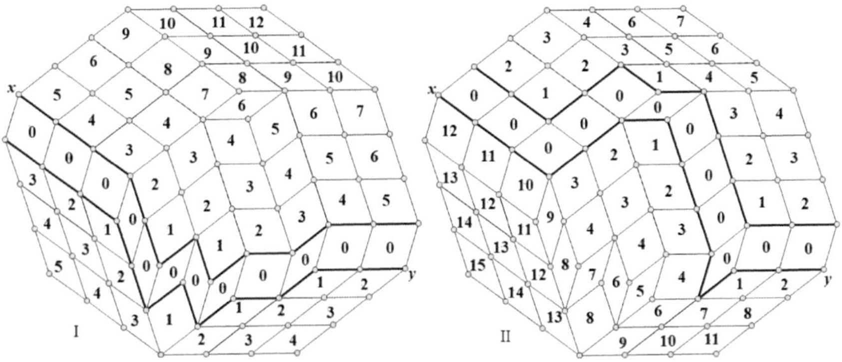

Figure 4.6

The free generating set in Figure 4.7(I) was obtained by interchanging some 0-tiles with 1-tiles in Figure 4.6(I). In Figure 4.7(II) the indicated s-1 tiles do not form a free generating set. The reason is that no tile in the indicated tile stream can be generated.

From this observation we can conclude that a free generating set for $\mathbf{G}_P^{\,d}$ *must contain at least one tile from each stream in* G_P*.*

Theorem 4.4 *Let P be a semiregular polygon with semiperimeter s. For any tiling of P by rhombs, there exists a free generating set for* $\mathbf{G}_P^{\,d}$ *with s-1 vertices.*

Proof. Let $\rho = (G_1, G_2, \ldots, G_{s-1})$ denote a path in $\mathbf{G}_P^{\,d}$ of length s-1 where G_1, G_{s-1} are points which correspond to tiles which touch opposite corner points of P. (See the example in Figure 4.8). We will argue that the vertices in ρ constitute a free generating set for $\mathbf{G}_P^{\,d}$. Let Γ denote $\{G_1, G_2, \ldots, G_{s-1}\}$ in union with any point of G_P generated by these points. We use the set Γ to accumulate generated points. Our aim

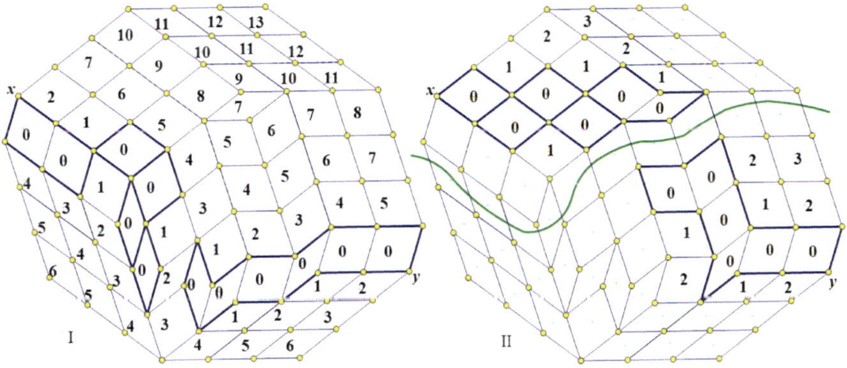

Figure 4.7

will be to show that Γ ultimately contains every vertex of $\mathbf{G}_P^{\,d}$ with no *redundancy*. That is to say that every point in $\mathbf{G}_P^{\,d}$ is generated by exactly two previously generated points, (which may include one or two vertices of ρ).

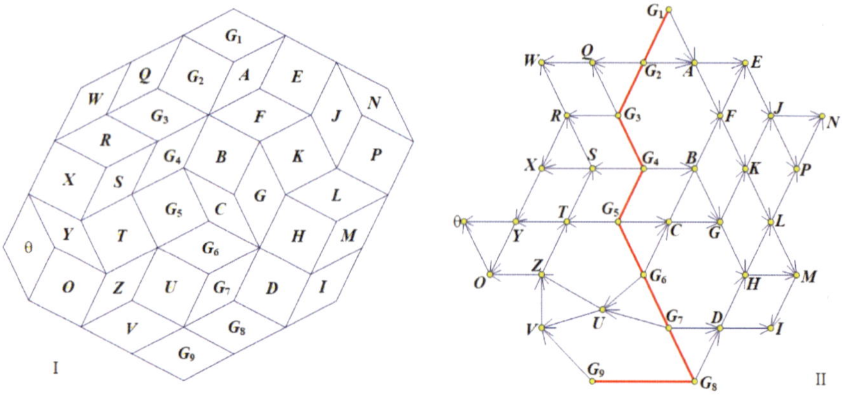

Figure 4.8

The path ρ divides $\mathbf{G}_P{}^d$ into a *left* section (L) and a *right* section (R). We will show that the points in ρ generate all the points in either section. Let R_1 denote the set of points in R which are adjacent to ρ. In our example in Figure 4.1, these are {A,B,C,D}. The set R_1 is necessarily smaller than the number of points in ρ and there are more edges from ρ to R_1 than there are points in R_1. As a consequence, at least one point in R_1 must be incident with two points of ρ. This point is thus generated by the points in Γ and is added to Γ. In our example there are 7 edges joining ρ and R_1. Since R_1 contains only 4 points, there must be 3 points, {A,C,D}, in R_1 which are generated by Γ.

The argument is then repeated using the new boundary of Γ. In our example in Figure 4.8(II), this boundary consists of the points {A,B,C,D}. There are necessarily more arcs out of points in this boundary to adjacent points not yet included in Γ. As a consequence at least one of these adjacent points must receive two arcs from points in Γ, and thus be included in Γ. Ultimately every point in R will be included in Γ, and a similar argument will result in every point in L being included in Γ. □

Similar to Theorem 4.2, we have the following result pertaining to $\mathbf{G}_P{}^d$.

Theorem 4.5 *Let P be a semiregular polygon with semiperimeter s. For any tiling of P by rhombs, every free generating set for* $\mathbf{G}_P^d$ *has exactly s-1 vertices.*

Proof. Let r denote the number of tiles used to cover P, and Let S be a free generating set for $\mathbf{G}_P^d$ consisting of x points. The parameter r also represents the number of vertices of $\mathbf{G}_P^d$. Assign variables to every vertex of $\mathbf{G}_P^d$ except those which lie in S. Thus the total number of variables is r - x.

A locally nilpotent assignment to the vertices of $\mathbf{G}_P^d$ requires a 0-sum equation for each interior point of $\mathbf{G}_P$. The number i of interior points of $\mathbf{G}_P$ is given by i = v - 2s, where v denotes the total number of vertices of $\mathbf{G}_P$ and 2s is the number of boundary points. To obtain a unque solution for each variable, the number of equations (v-2s) is equated with the number of variables (r-x) to obtain

(1) $x = r - v + 2s$.

From Theorem 4.2 we have

(2) $e - 2r = s$,

and Eulers formula [4] yields

(3) $r = e - v + 1$.

From (2), (3) we obtain

(4) $0 = s + r - v + 1$,

and finally (1), (4) yield $x = s-1$. □

4.3 Exercises

(1) The polygon P in Figure 4.9 is the same as the one given in Figure 4.1, and also has exactly the same tiling. For the indicated diametric set determine its *efficiency*, i.e., how many iterations are required to generate P? Which diametric set in P is the most efficient? A diametric set ρ creates a partition of P into two subpolygons P_1, P_2. Is ρ most efficient when the order of P_1 and the order of P_2 are equal or nearly the same?

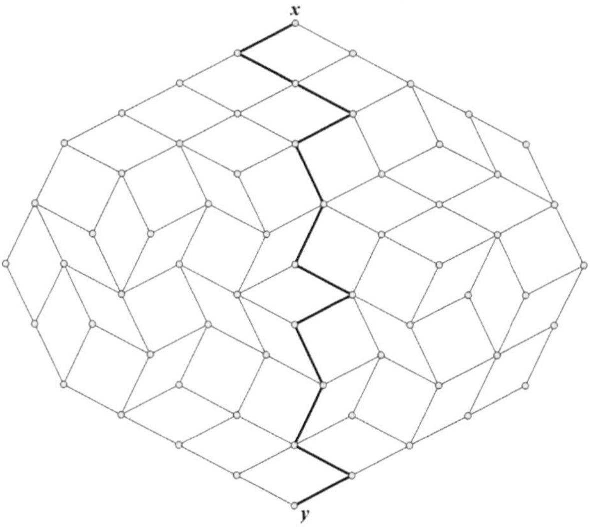

Figure 4.9

(2) Among all the rhomb tilings of a given semiregular polygon P, which has the most efficient generation?

(3) For the semiregular polygon in Figure 4.8 which corner to corner path ρ is the most efficient generator for $G_P{}^d$? For a given rhomb tiling of a semiregular polygon which corner to corner path is the most efficient generator of $G_P{}^d$? Among all the rhomb tilings of a given semiregular polygon P, which has the most efficient generation for $G_P{}^d$?

CHAPTER 5

Free Generating Sets in Flowers

5.1 Introduction

The free generating sets described in Chapters 3, 4 are *simple* in the sense that there is a clear path to follow for the generation: a series of point extensions as described in Section 3.3 are used. In this Chapter free generating sets are constructed for which it is not immediately clear that *any* points are generated. An example of such a generating set is given in Figure 5.1.

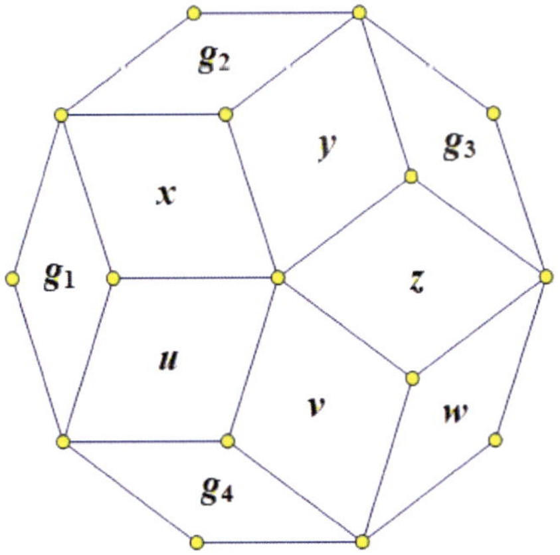

Figure 5.1

A locally nilpotent assignment for the decagon in Figure 5.1 would entail the satisfaction of the equations

(1) $g_1 + x + u = 0$
(2) $g_2 + x + y = 0$
(3) $g_3 + y + z = 0$
(4) $g_4 + u + v = 0$
(5) $v + z + w = 0$
(6) $x + y + z + u + v = 0$.

Using equations (2), (4), (6) we obtain $z = g_2 + g_4$. The values for the remaining variables are then easily determined: $v = g_1 + g_3$; $x = g_3 + g_4$; $y = -g_2 - g_3 - g_4$; $u = -g_1 - g_3 - g_4$; $w = -g_1 - g_2 - g_3 - g_4$. All six equations are satisfied by these values and thus $\{g_1, g_2, g_3, g_4\}$ is a free generating set. This type of generating set we call *obscure.*

We call the tiling in Figure 5.1 a *flower.* In general the *flower* F_n is a regular polygon with semiperimeter n which features point extensions of a central *star* consisting of n rhombs. The point extensions keep the n-fold rotational symmetry of the star and form *rings* around the star. In Figure 5.2, the flower F_{10} is illustrated. The central star is the first ring $R_1 = \{x_1, \dots, x_{10}\}$, then the second ring is $R_2 = \{y_1, \dots, y_{10}\}$, etc. There are a total of 4 rings with R_4 consisting of the 10 boundary tiles. Whenever n is even, F_n has n sides each with two unit edges.

In Figure 5.3 the flower F_{11} is illustrated. Instead of attempting a difficult construction with rhombs, lattice parallelograms are used. This results in a distortion of the shapes of tiles, but is nevertheless adequate for our analysis.

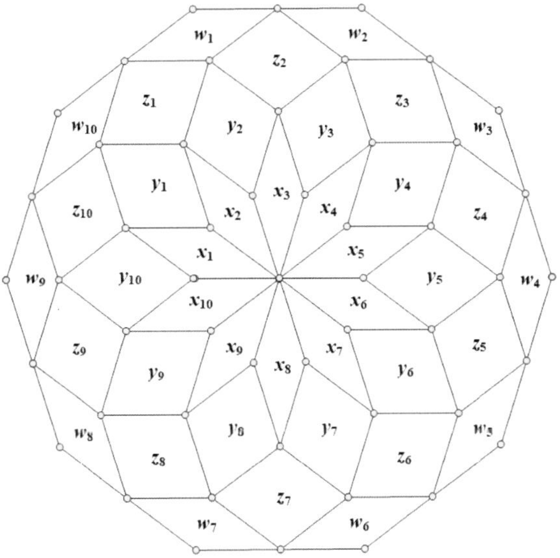

Figure 5.2

If n is odd then F_n will be a regular polygon with 2n sides. Clearly the central star, R_1 is a simple free generating set, but as in Figure 5.1 where we have shown that R_2 is an (obscure) free generating set, some of the other rings also provide examples of obscure free generating sets.

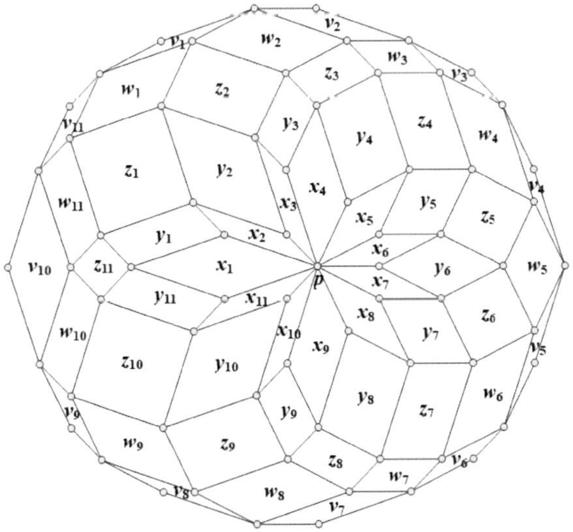

Figure 5.3

To determine which rings form obscure free generating sets in F_n it is useful to establish some properties of polar tilings of regular polygons.

Theorem 5.1 *Let P be a simple regular polygon with semiperimeter s, and with a polar tiling with rhombs. Let R denote the tile set of P and let* $\phi: Z \dashrightarrow R$ *be a locally nilpotent assignment. Let* $X = \{x_1, \ldots, x_{s-1}\}$ *denote the set of integers assigned to the polar rhombs and let T denote the value assigned to the top tile. If s is even then* $T = \Sigma X$*, and if s is odd then* $T = -\Sigma X$*.*

Proof. The contention of the Theorem is easily seen to be true in case s = 3,4. Assume inductively that the Theorem holds for every simple regular polygon with a polar tiling and semiperimeter less than s. Let a, b, c denote the integer values assigned to the tiles adjacent to the top tile as in Figure 5.4. Let P_a, P_b, P_c denote subpolygons of P with top tiles a, b, c respectively. These inherit polar tilings from the polar tiling of P and all have the same pole as P. The semiperimeters of P_a, P_b, P_c are s-1, s-1 and s-2 respectively.

Suppose s is even, then s-1 is odd and by the inductive hypothesis $a = -(\Sigma X - x_{s-1})$; $b = -(\Sigma X - x_1)$, and $c = (\Sigma X) - x_1 - x_{s-1}$. Thus $T = -a - b - c = (\Sigma X - x_{s-1}) + (\Sigma X - x_1) - \Sigma X + x_1 + x_{s-1} = \Sigma X$. If s is odd then s-1 is even and s-2 is odd, so $a = \Sigma X - x_{s-1}$; $b = \Sigma X - x_1$ and $c = -(\Sigma X) + x_1 + x_{s-1}$, and consequently $T = -a - b - c = -(\Sigma X - x_{s-1}) - (\Sigma X - x_1) + (\Sigma X) + x_1 + x_{s-1} = -(\Sigma X)$. □

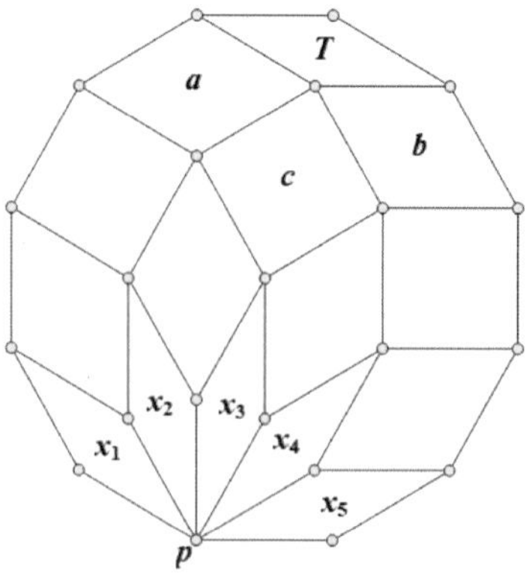

Figure 5.4

Of course $\Sigma R_1 = 0$ as per the requirement of a locally nilpotent assignment. In fact, if R is any ring in a flower tiling with a nilpotent assignment, then $\Sigma R = 0$. We leave the proof as an exercise.

Theorem 5.2 *Let the flower F_n have a locally nilpotent assignment. If R is any ring in F_n then $\Sigma R = 0$.* □

If R is a ring in F_n, let R^* denote the set of values in R with one value deleted. The values in R_1 clearly generate F_n but strictly speaking R_1 is *not* a free generating set. This is due to the result in Theorem 5.2: since $\Sigma R = 0$, one of the values in R is dependent on the choices for the remaining values. All but one of the values in R_1 can be assigned freely, so we can say most properly that R_1^* is a free generating set. Our next objective will be to show that certain of the *deleted* rings, (R_t^*) in F_n *are not* free generating sets. To that end the following Lemma will prove useful.

Lemma 5.3 *Let $S = (x_1, x_2, \dots, x_n)$ be a sequence of integers with $\Sigma S = 0$, and let v denote a consecutive subsequence of S with $m<n$ elements. Let $d = gcd(m,n)$ and let $\phi_d(v)$ denote a left circular shift of v by d units. Then*

$$\Sigma v + \Sigma\, \phi_d(v) + \Sigma\, \phi_{2d}(v) + \dots + \Sigma\, \phi_{(m-1)d}(v) = 0.$$

Proof. Let the matrix M be defined as in Figure 5.5.

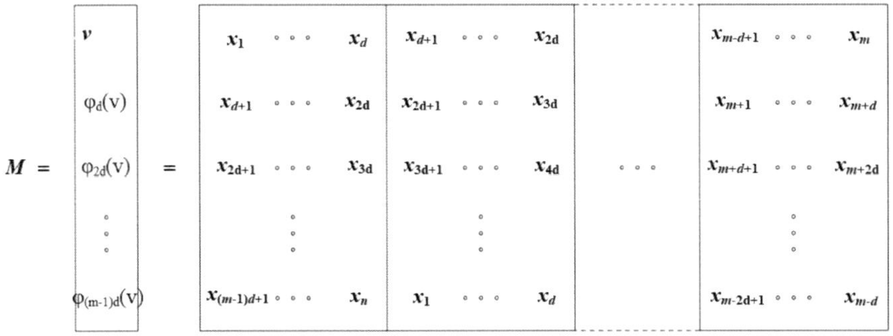

Figure 5.5

Note that M consists of the indicated m/d blocks. Each block is obtained from its predecessor by a circular shift of the rows. The first block contains all n of the elements of S, and thus each block contains precisely all the elements of S. Since $\Sigma S = 0$, it follows that $\Sigma M = 0$. □

Theorem 5.4 *The deleted ring R_m^* is a free generating set for the flower F_n iff gcd(m,n) = 1.*

Proof. Let $S = (x_1, \dots, x_n)$ denote the sequence of integer values assigned to the central star of F_n in any locally nilpotent integer assignment for F_n. Let d = gcd(m,n) and let ϕ_d denote the automorphism of F_n which rotates the central star by d units clockwise. Let $v = (x_1, \dots, x_m)$ and let P(z) denote the polar subpolygon of F_n whose polar rhombs have the values in v, and whose top value z lies in R_m.

By Theorem 5.1, $\Sigma v = z$ if m is odd and $\Sigma v = -z$ if m is even. By Lemma 5.3

$$\Sigma v + \Sigma\phi_d(v) + \Sigma\phi_{2d}(v) + \dots + \Sigma\phi_{(m-1)d}(v) = 0.$$

The top value of the polar subpolygon with the values in $\phi_d(v)$ is $\phi_d(z)$, and in general, the top value of the polar subpolygon with the values in $\phi_{kd}(v)$ is $\phi_{kd}(z)$. So if m is odd then $0 = \Sigma v + \Sigma\phi_d(v) + \Sigma\phi_{2d}(v) + \dots + \Sigma\phi_{(m-1)d}(v) = z + \phi_d(z) + \dots + \phi_{(m-1)d}(z)$. If $d \geq 2$ then $Q = \{z, \phi_d(z), \dots, \phi_{(m-1)d}(z)\}$ has fewer than m elements, and since $\Sigma Q = 0$, these elements cannot constitute a free generating set for F_n. A similar conclusion follows in case m is even. Note that if d = 1 then Q contains every element in R_m and $\Sigma Q = 0$ is just the result of Theorem 5.2. It then remains open in this case as to whether R_m^* is a free generating set for F_n.

Now suppose gcd(m,n) = 1, and consider first the case in which m is odd. Let $R_m = (z_1, z_2, \dots, z_n)$. For each $i \in \{1, \dots, n\}$, let P_i denote the polar subpolygon of F_n with top value z_i and polar rhombs $X_i = \{x_i, x_{i+1}, \dots, x_{i+m-1}\}$. By Theorem 5.1 $z_i = \Sigma X_i$ for each i. Since gcd(m,n) = 1, there exists a positive integer s such that sm = 1 (mod n). Since sm-1 is a multiple of n, and $\Sigma\{x_i : i = 1, \dots , n\} = 0$, the sum of sm consecutive values x_i, taken clockwise around the central star is just equal to exactly one of the values assigned to a rhomb in the central star.

When the sm consecutive values are divided into s sequences each with m elements, we obtain the sum of s values in the ring R_m. Thus the sum of s values in R_m equals just one of the values assigned to a rhomb in the central star. By symmetry each of the central rhomb values can be obtained in this way. Since the central rhomb values generate the nilpotent assignment for the entire flower F_n, it follows that The values in R_m also generate F_n. Since $\Sigma R_m = 0$, it follows that R_m^* is an obscure free generating set for F_n.

In case m is even, the argument is similar so we omit the details. □

5.2 Exercises

(1) If R is any ring in a flower tiling with a locally nilpotent assignment, show that $\Sigma R = 0$.

(2) For the flower F_{10} in Figure 5.2, supply integer values for the rings R_3, R_4 with $\Sigma R_3 = 0$ and $\Sigma R_4 = 0$, then check that R_3^* is a free generating set, but not R_4^*.

(3) Show that if n is odd, F_n is a regular polygon with 2n sides each with length 1 unit, and if n is even, F_n has n sides each with length 2 units.

CHAPTER 6

Free Generating Sets in Rectangles

6.1 Preliminaries

Our purpose in this Chapter will be to describe all free generating sets in rectangular polysquares. Let P be a rectangular polysquare with tile set T, and let $\phi:T \to Z$ be a locally nilpotent assignment of integers to the tiles, (i.e., unit squares or cells) of P. Every interior point of G_P has a nbd consisting of four squares whose assigned values must sum to 0. In case both dimensions of P are even then P has a nbd dissection and thus is nilpotent.

If P has semiperimeter s then a free generating set, (FGS) has exactly s-1 tiles, consistent with Theorem 4.4. Many simple FGS's can be constructed which consist of tile paths of length s-1 or *crosses* or *trees* with s-1 cells as illustrated in Figure 6.1.

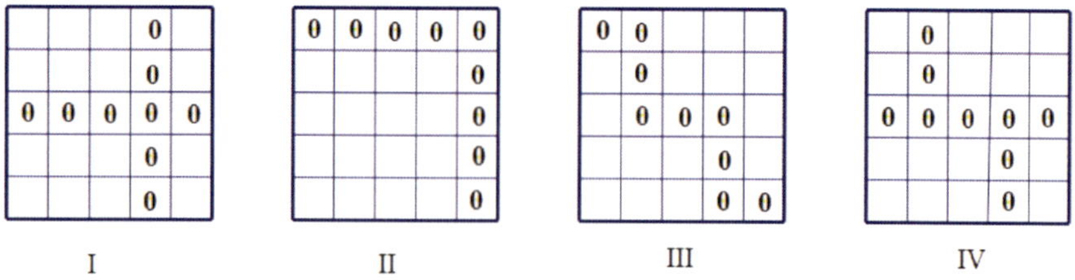

Figure 6.1

Besides the requirement of having exactly s-1 cells, the following propeties of an FGS for a polysquare P are easily seen.

(I) An FGS must contain at least one cell from each row and column of P.

(II) If a cell of an FGS occupies the intersection of row A and Column B of P, then one of row A, Column B must contain at least one more cell belonging to the FGS. That is to say that no cell belonging to an FGS can be a *singleton* in both its row and column.

(III) No column (row) of P can consist entirely of elements of an FGS which are singletons in their respective rows (columns).

If a subset S of T contains no cells in some row (column) then none of the elements in that row (column) can be generated, and similarly, in case S contains a cell which is alone in its row and column, then the remaining elements in its row and column cannot be generated.

A fourth property:

(IV) An FGS cannot contain all four corner cells of any subrectangle of P,

is a consequence of the following Theorem.

Theorem 6.1 *Let P be a rectangular polysquare with a locally nilpotent assignment, and let A, B, C, D be the integer values assigned to the four corners of a subrectangle R of P, as in Figure 6.2. (i) If all sides of R have odd length then A+D = B+C, i.e., the sum of diagonally opposite corner values is the same; (ii) if all sides of R are even then A+B+C+D = 0; (iii) if one side, say AB is even and the adjacent side AC of R is odd then A+B = C+D.*

Proof. In case P,Q,R,S are integer values assigned to the four corners of a subrectangle R, we will use R = [P,Q,R,S] to denote the rectangle, where P is the top and Q the bottom value of the first column, and R is the top and S the bottom value of the last column of R. Figure 6.2(I) represents a rectangle with both dimensions odd.

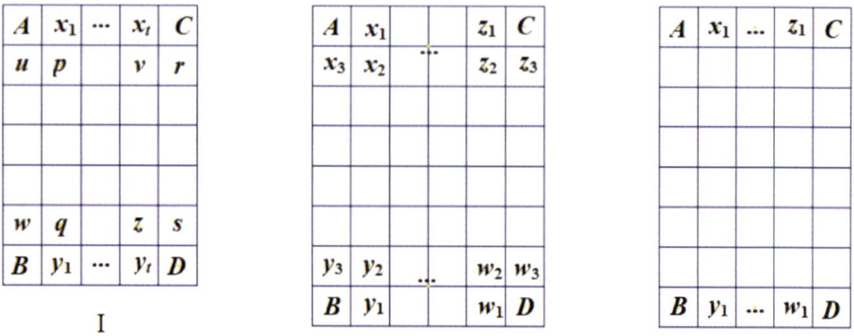

I II III

Figure 6.2

Let $R_1 = [A, w, x_t, z]$; $R_2 = [u, B, v, y_t]$; $R_3 = [x_1, q, C, s]$; $R_4 = [p, y_1, r, D]$, and let $X = x_1 + ... + x_t$ and $Y = y_1 + ... + y_t$. These rectangles have both dimensions even and thus are nilpotent, i.e., $\Sigma R_1 = \Sigma R_2 = \Sigma R_3 = \Sigma R_4$. Then

(1) $0 = \Sigma R_1 - \Sigma R_2 = (A + X) - (B + Y)$, and
(2) $0 = \Sigma R_3 - \Sigma R_4 = (X + C) - (Y + D)$.

Thus $A + X - B - Y = X + C - Y - D$, which yields $A - B = C - D$ or $A + D = B + C$.

Now suppose both dimensions of R are even as illustrated in Figure 6.2(II). If R is 2 by 2 then $A+B+C+D = 0$ holds immediately due to the locally nilpotent assignment. It is also easily seen that $A+B+C+D = 0$ in case R is 2 by 4, 2 by 6, 2 by 8 or 4 by 4. Suppose R has area greater than 16, and assume inductively that all rectangles with two even dimensions and area less than R satisfy the contention of the Theorem. In Figure 6.2(II) subrectangles $R_1 = [x_1, y_1, z_1, w_1]$, $R_2 = [x_2, y_2, z_2, w_2]$, $R_3 = [x_3, y_3, z_3, w_3]$ all have two even dimensions and area less than R. By the inductive hypothesis

$x_1 + y_1 + z_1 + w_1 = 0$; $x_2 + y_2 + z_2 + w_2 = 0$ and $x_3 + y_3 + z_3 + w_3 = 0$.

Adding these equations yields

$$(x_1 + x_2 + x_3) + (y_1 + y_2 + y_3) + (z_1 + z_2 + z_3) + (w_1 + w_2 + w_3) = 0, \text{ and thus}$$

$$(-A) + (-B) + (-C) + (-D) = 0, \text{ or } A+B+C+D = 0.$$

Finally suppose Case (iii): side AB is even and side AC is odd, as illustrated in Figure 6.2(III). Define subrectangles $R_1 = [A, B, z_1, w_1]$, $R_2 = [A, B, x_1, y_1]$, $R_3 = [x_1, y_1, C, D]$, and $R_4 = [z_1, w_1, C, D]$. These rectangles all have two even dimensions, so by Case (ii)

(3) $A + B + z_1 + w_1 = 0$;
(4) $A + B + x_1 + y_1 = 0$;
(5) $x_1 + y_1 + C + D = 0$;
(6) $z_1 + w_1 + C + D = 0$.

Subtracting (5) from (3) yields

(7) $(A+B) - (C+D) + (z_1 + w_1) - (x_1 + y_1) = 0$,

And subtracting (5) from (6) yields

(8) $(z_1 + w_1) - (x_1 + y_1) = 0$.

Combining (7), (8) yields $(A+B) - (C+D) = 0$, or $A+B = C+D$ as stipulated. □

Theorem 6.1 shows that if P is a rectangular polysquare with a locally nilpotent assignment, then the four corners of any subrectangle have a specific relationship, a *corner equation,* and their values cannot be freely chosen. Property (IV) is thus a consequence.

6.2 L-inversions

Properties (I)-(IV) of section 6.1 are not yet sufficient to determine an FGS in a rectangular polysquare. Consider the example in Figure 6.3 which satisfies (I) - (IV). Applying Case (ii) of Theorem 6.1 we obtain

(1) A + x = B + C, and
(2) F + x = E + D.

These equations lead to the relation B + C - A = E + D - F. So the cells A, B, C, D, E, F cannot belong to an FGS. In Figure 6.3 we have indicated a graph with vertices marking cells in the set S = {A,B,C,D,E,F,G,H,I,J} and with edges joining elements in S which are adjacent in either a row or column. We denote this graph by Γ(S). Evidently the existence of the cycle (A, B, E, F, D, C) has caused the failure of S to be an FGS. In this Section we will establish the following additional requirement for an FGS in a rectangular polysquare.

(V) If S is an FGS in a rectangular polysquare then the graph Γ(S) is acyclic.

Note that (V) is a generalization of property (IV).

To assist in the demonstration of property (V) we introduce the notion of an *L-inversion*. Let P be a rectangular polysquare and S an FGS in P. If R is a subrectangle of P three corners of which lie in S, (thus forming an L-shaped subset of S), then if any one of the three corner cells in S is removed and the cell occupying the vacant corner of R is inserted in S, (thus recreating an L shape), we call this transformation of S an *L-inversion*. In the following Theorem we show that L-inversion transforms an FGS into another FGS.

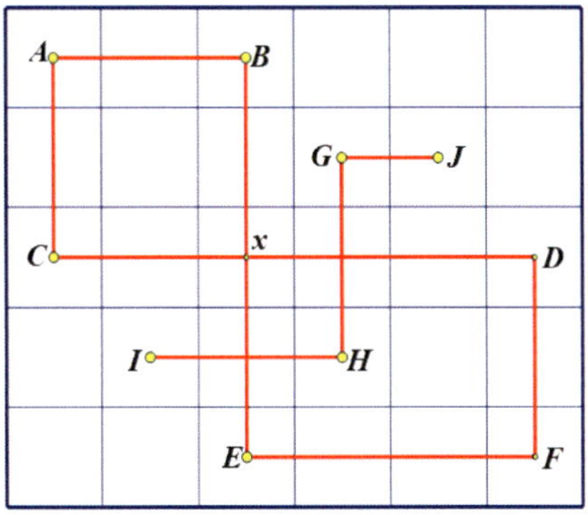

Figure 6.3

Theorem 6.2 *Let P be a rectangular polysquare and let S be an FGS in P. Then any L-inversion applied to S produces a set S′ which is also an FGS.*

Proof. Let T denote the set of cells which comprise P and let $\phi:T\to Z$ denote a locally nilpotent assignment. Let R be a subrectangle of P with corner cells A, B, C, D as in Figure 6.4. Suppose cells A, B, C lie in S thus forming an L-shaped subset of S, and let $S' = (S - B) \cup D$. The transformation $S\to S'$ is thus an L-inversion. It is clear from Theorem 6.1 that the values $\{\phi(A), \phi(C), \phi(D)\}$ belonging to the cells A, C, D of S′ will generate the value $\phi(B)$ and thus that the values belonging to cells in S′ will generate all the values generated by S, i.e., the entire locally nilpotent assignment.

It remains to argue that S′ is an FGS. To this end it suffices to show that no *nontrivial relation* exists among the set of values $\{\phi(x) : x \in S'\}$. (By *nontrivial relation* we mean to exclude obvious identities such as $g_1 + g_2 - g_3 = g_2 - g_3 + g_1$.) Suppose, to the contrary that there exists a nontrivial relation among the set of values $\{\phi(x) : x \in S'\}$. If $S = \{A, B, C, X_1, \ldots, X_n\}$ then $S' = \{A, C, D, X_1, \ldots, X_n\}$ and a relation among the values belonging to cells in S′ has the form

(1) $\varepsilon\phi(A) + \delta\phi(C) + \gamma\phi(D) + \varepsilon_1\phi(X_1) + \ldots + \varepsilon_n\phi(X_n) = 0,$

Where $\varepsilon, \delta, \gamma, \varepsilon_1, \ldots, \varepsilon_n$ are integers. Now by Theorem 6.1 we have one of the following

(2) $\phi(D) = -\phi(A) - \phi(B) - \phi(C)$,
(3) $\phi(D) = \phi(A) + \phi(B) - \phi(C)$,
(4) $\phi(D) = \phi(A) + \phi(C) - \phi(B)$,
(5) $\phi(D) = \phi(B) + \phi(C) - \phi(D)$.

Now substituting for $\phi(D)$ in (1) we obtain a nontrivial relation among the values $\{\phi(x) : x \in S\}$. But this contradicts the character of S as an FGS. □

In Figure 6.4 a sequence of L-inversions are indicated. In 6.4(I) an L shaped arrangement of free generators is illustrated. It is an example of a simple FGS. From this arrangement a series of L-inversions are outlined arriving at the construction in 6.4(VIII) which must, by Theorem 6.2, be an FGS. At each step in the series the graph Γ(S) is drawn. This graph begins as a tree and remains a tree throughout. (A *tree* is a connected acyclic graph.) Our next objective will be to show that if S is a collection of s-1 cells in a rectangular polysquare with semiperimeter s, and Γ(S) is a tree, then S is an FGS.

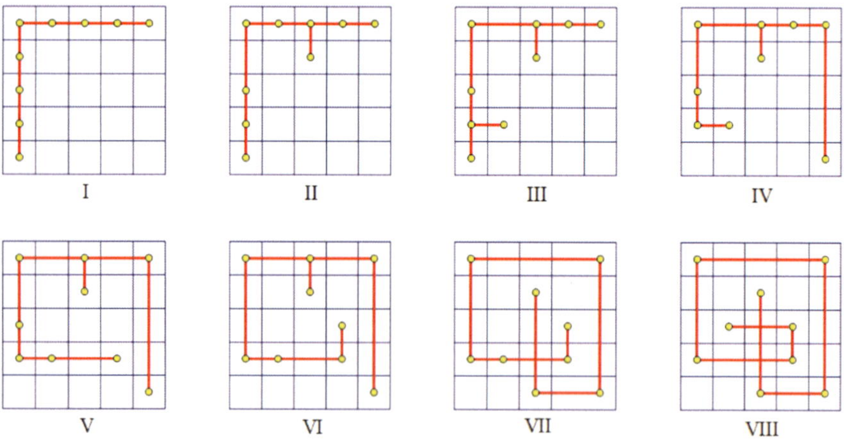

Figure 6.4

Theorem 6.3 *Let P be a rectangular polysquare with semiperimeter s and let S be a subset of the cells of P with s-1 elements. If the graph Γ(S) is a tree, then S is an FGS.*

Proof. Since Γ(S) has s-1 elements it must contain at least one cell in each row and column of P. Let C denote the column of P which contains a cell X of S in its first row. We will use L-inversions to move cells of S either to the top row or to column C. Starting with cell X we traverse the tree Γ(S) and use L-inversions in a systematic fashion, to move cells either to the top row or to column C. It makes little difference to our argument which direction we proceed from cell X along the tree, so let us just consider the case in which we move down to cell Y and ultimately to the right say to cell 1, as in the example in Figure 6.5(I).

L-inversions can be used to move every cell of S in line with X, Z to the top row. (These are cells 1,2,3 in Figure 6.5(I).) When the tree makes a right angle, then it will be possible to use L-inversions to move cells of S to column C. (Cells 4,5 in Figure 6.5 can be moved to column C.) Whenever a right angle occurs in the evolving tree, we change from moving to column C to moving to the top row or from the top row to column C. In Figure 6.5(I) the cells destined for the top row are indicated in blue, and the cells destined for column C are indicated in green. Red is used to mark the cells of S which already lie in column C.

When the full series of L-inversions are accomplished a configuration consisting entirely of cells in the top row and column C is obtained. This configuration is easily seen to be a simple FGS, and by Theorem 6.2 it follows that the original configuration S is also an FGS. □

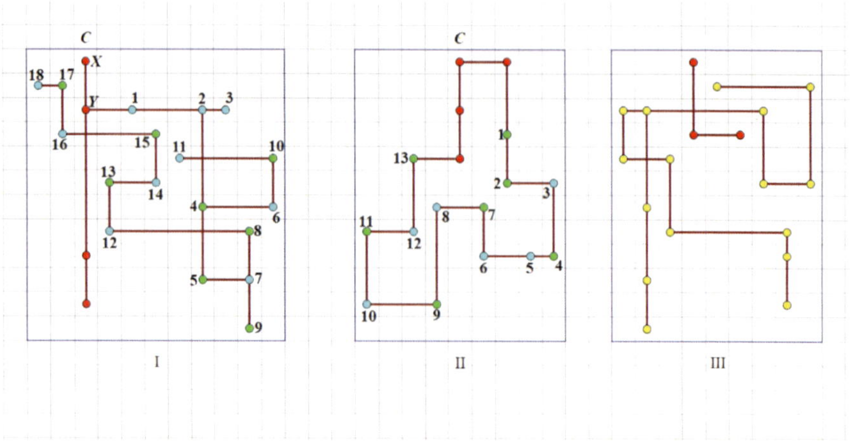

Figure 6.5

It remains to show that if S is an FGS then Γ(S) must be a tree. Suppose, as in Figure 6.5(II), that Γ(S) contains a cycle. The same color scheme is used to indicate the movements caused by L-inversions: green elements are moved to column C and blue elements are moved to the top row. Elements already occupying the top row or column C are red. There is always a problem in allocating the last non-red member of the cycle. In Figure 6.5(II) it is element number 13; it is due to be shifted by L-inversion to column C, but its destination is already occupied by a red element.

Finally we argue that Γ(S) must be connected. The basic observation is that if we count together the rows and columns that the cells a tree with m elements occupies, then we obtain a total of m+1. This is because the first cell chosen occupies one row and one column, and adding one neighbor at a time to the tree adds exactly one to the total count of occupied rows and columns. Suppose Γ(S) is not connected and let T denote a connected component of Γ(S) which contains m generators with m < s-1. T is a tree so the cells in T occupy a total of m+1 rows and columns. This leaves a total of s-m-1 free rows and columns which must be occupied by the remaining s-m generators. These s-m generators also must not form any cycle. But by our first observation, these elements need a total of s-m+1 rows and columns.

In Figure 6.5(III) T is represented by the red dotted cells. It contains m = 3 elements and occupies a total of m+1 = 4 rows and columns. Since s = 21 there are a total of s-4 = 17 remaining rows and columns to be occupied by the remaining s-m = 18 points. T represents a component of Γ(S) which is not connected to any of the remaining points of S, so the 18 points cannot occupy any of the 4 rows and columns used by elements of T. Also the 18 points are not allowed to form any cycle since that possibility has been eliminated in the previous paragraph. The 18 points then require 19 free rows and columns but only 17 are available. The yellow dots in 6.5(III) indicate the maximum number of cells, (only 16), which can be placed in the rectangle without creating a cycle or joining T.

We have thus completed the proof of the following main result of this Chapter.

Theorem 6.4 *Let P be a rectangular polysquare with semiperimeter s and let S be a subset of the cells of P. Then S is an FGS iff S has exactly s-1 elements, and Γ(S) is a tree.* □

6.3 Exercises

1. In Figure 6.6(I) a polysquare P with semiperimeter s = 21 is illustrated. The red squares indicate the locations of an FGS S. Note that S contains s-1 = 20 cells and that Γ(S) is a tree, as required by Theorem 6.4. Choose any 20 integers to occupy the cells belonging to the FGS and use these to generate a locally nilpotent assignment for P.

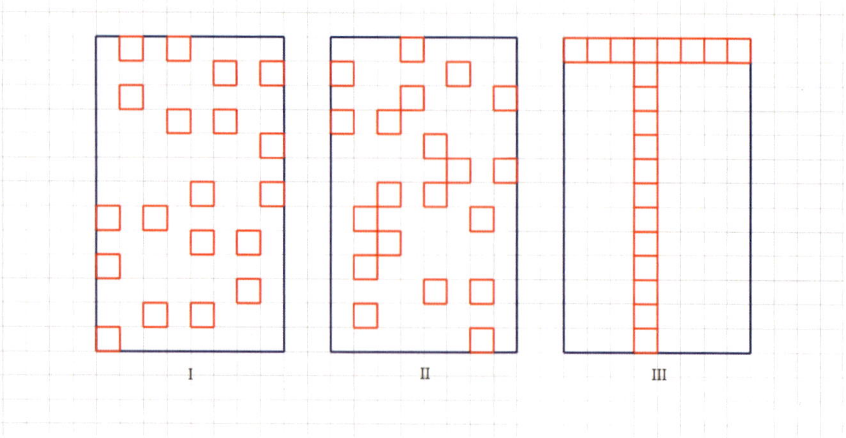

Figure 6.6

2. In Figure 6.6(II) an FGS S is given in a rectangular polysquare with semiperimeter 21. Use L-inversions to transform S to the FGS S′ in Figure 6.6(III).

3. Let S be an FGS in a rectangular polysquare P. If two rows or two columns of P are interchanged a new set of cells S′ is obtained. Is S′ also an FGS?

4. Let S denote a collection of cells in a rectangular polysquare P and suppose Γ(S) is a cycle as in Figure 6.5(II). Show that there exists a sequence of L-inversions which results in a set S′ of cells all of which lie on the four sides of a rectangle with an element of S′ occupying each corner cell of the rectangle. This is another argument that can be used to show that if S is an FGS then Γ(S) must be acyclic.

CHAPTER 7

Stream Polygons

7.1 Introduction

Let P be a polygon tiled with rhombs, (or parallelograms). If each side of P is a tile stream then we call P a *stream polygon*. A stream triangle is illustrated in Figure 7.1 as a subpolygon of a regular octagon with a locally nilpotent assignment. The three corner values are circled. We will show that the corner values of any stream polygon satisfy a relationship which depends on the lengths of the bounding tile streams.

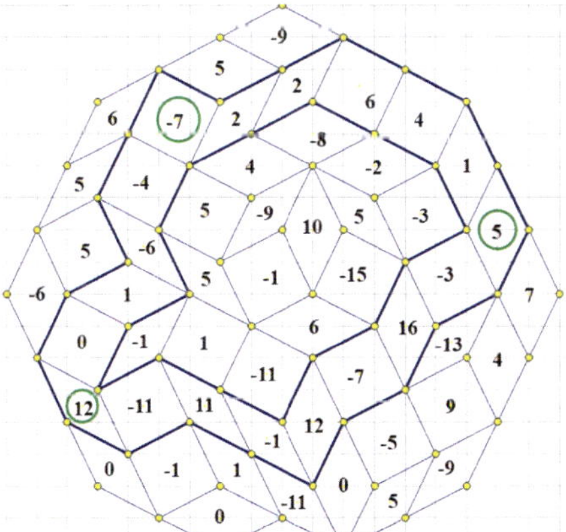

Figure 7.1

In Figure 7.2 a stream quadrangle is indicated in the same octagon and with the same locally nilpotent assignment as in Figure 7.1. The corner values are again circled.

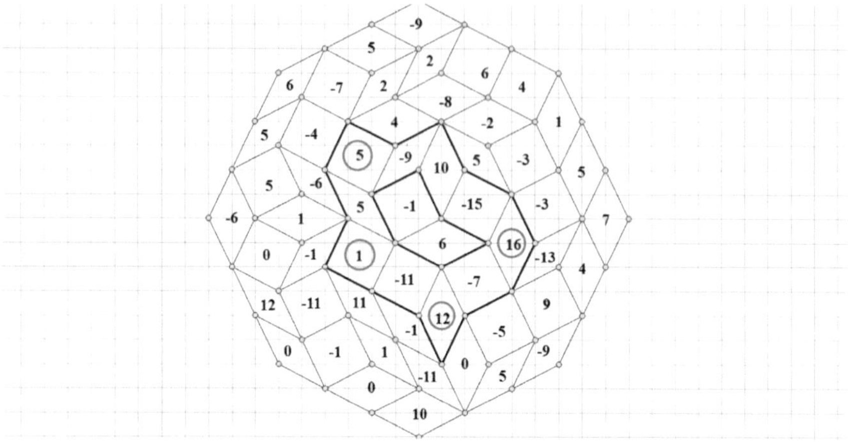

Figure 7.2

For the purpose of analysis, the stream triangle in Figure 7.1 will be represented by the schematic diagram in Figure 7.3(I), and the stream quadrangle will be represented by the diagram in Figure 7.3(II). For the triangle corners A, B, C are the tiles with values 12, -7, 5 respectively, and for the quadrangle, letters A, B, C, D represent corner tiles in Figure 7.2 with values 5, 1, 12, 16 repectively. The edge labels indicate the length of the corresponding tile stream.

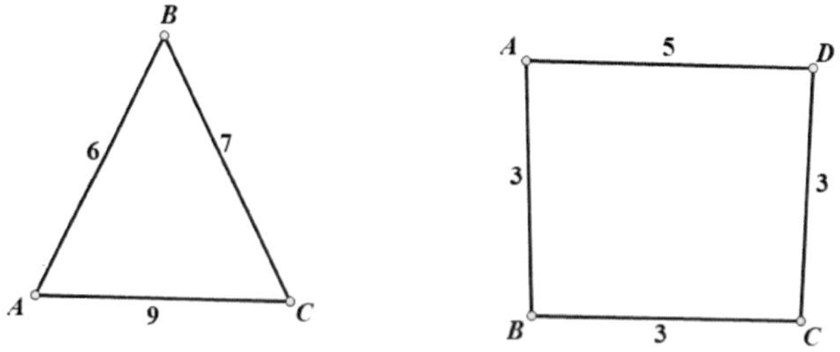

Figure 7.3

The sum of the tile stream lengths for any stream polygon P must be even since every stream which enters P at one boundary edge must also leave P at an opposite boundary edge. Thus stream triangles Fall into two types: (i) those with three even sides, and (ii) those with one even and two odd sides. Stream quadrangles fall into four types: (i) those with all four sides even; (ii) those with two adjacent sides even and two adjacent sides odd; (iii) those with one pair of opposite sides even and one pair of opposite sides odd; (iv) those with all four sides odd.

7.2 Stream Triangles and Quadrangles

In Figure 7.4 we have listed the different schematic diagrams for the types of stream triangles and quadrangles. With reference to this Figure we have the following result.

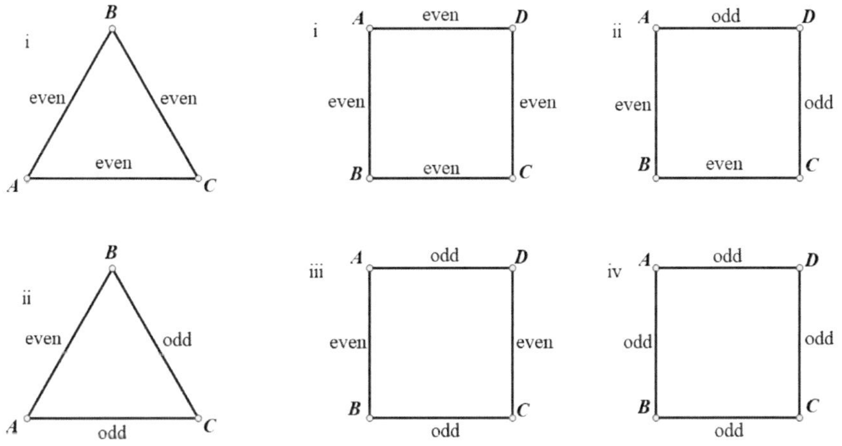

Figure 7.4

Theorem 7.1 *For any type (i) stream triangle A+B+C = 0; for any type(ii) stream triangle A+B = C; For any type(i) stream rectangle A+B+C+D = 0; for any type (ii) stream rectangle A+B+C = D; for any type(iii) stream rectangle A+B = C+D; for any type(iv) stream rectangle A+C = B+D.*

**Proof.* The argument will proceed inductively. As basis for the induction a number of small stream triangles and quadrangles need to be examined. These are listed in Figures 7.5, 7.6. We leave as an exercise to show that the Theorem stipulations hold for each of these small examples. Let Δ = ABC be a stream triangle and assume inductively that all stream triangles and quadrangles with smaller order than Δ satisfy the stipulations of the Theorem. (Note: the *order* of a polygon tiled with rhombs or parallelograms is the number of tiles used.)

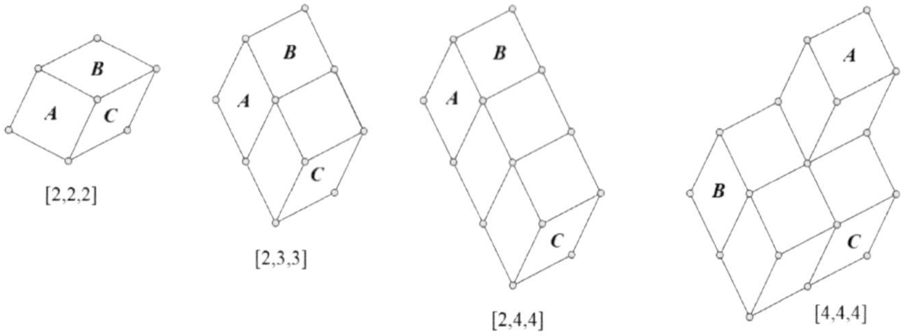

Figure 7.5

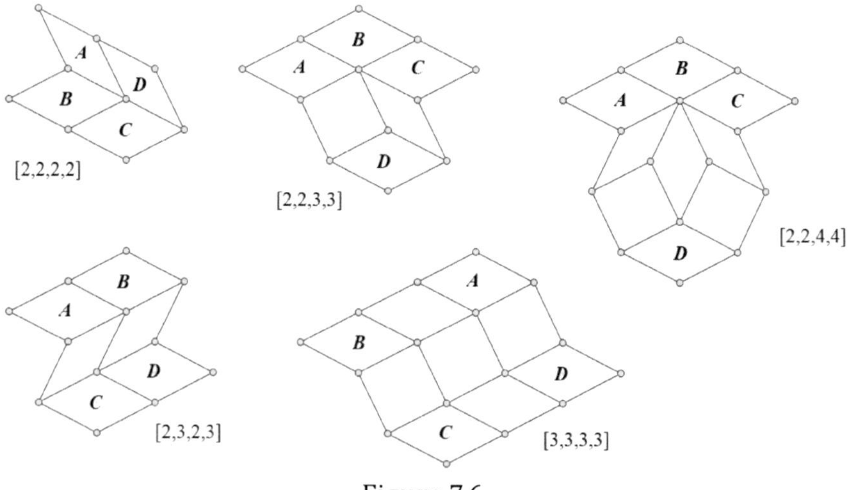

Figure 7.6

(i) We start by considering the type(i) stream triangle and construct a *cross-stream* ED as in Figure 7.7(I). Our intention is that E denotes a tile in stream AB and D denotes a tile in stream AC. We consider the following cases.

(1) *AD is odd and ED is odd.* Since AC is even for a type(i) stream triangle, it must be that DC is even. (Note: in this case odd + even = even since tile D is counted twice.) Now stream quadrangle EBCD has two even sides and one odd side. To maintain the required even perimeter, it must be that EB is even. Thus EBCD is a type(ii) stream quadrangle and thus (1) B+C+D = E by the inductive hypothesis. Since AB is even and EB is odd, it must be that AE is even and thus AED is a type(ii) stream triangle. Then(2) A+E = D by the inductive hypothesis. Combining (1),(2) yields A+B+C = 0.

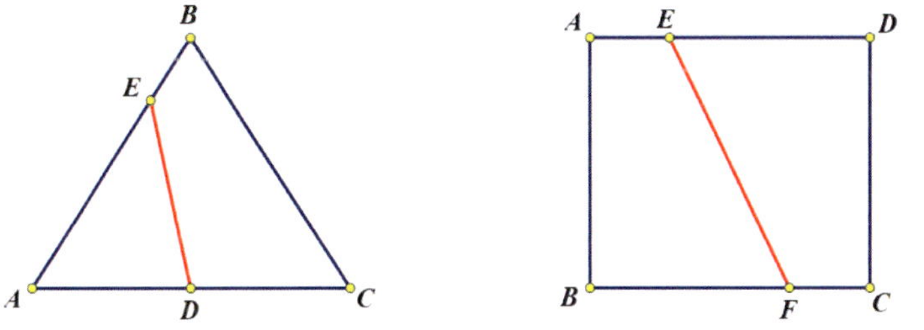

I II

Figure 7.7

(2) *AD is odd and ED is even.* In this case DC is even and stream quadrangle EBCD has three even sides. To maintain an even perimeter, EB must also be even making EBCD into a type(i) stream quadrangle. By the inductive hypothesis we have (3) E+B+C+D = 0. AE must be odd and thus AED is a type(ii) stream triangle. By the inductive hypothesis (4) E+D = A. Combining (3), (4) again yields A+B+C = 0.

(3) *AD is even and ED is odd.* In this case DC is odd; EB must be even and AE is odd. Stream quadrangle EBCD is type(ii) and stream triangle AED is type(ii). BY the inductive hypothesis we obtain (5) E+B+C = D and (6) A+D = E. Combining (5), (6) yields A+B+C = 0.

(4) *AD is even and ED is even.* In this case DC is odd; EB is odd and AE is even. Stream quadrangle is type(iii) and stream triangle AED is type(i). By the inductive hypothesis (7) E+D = B+C and (8) A+E+D = 0. Combining (7),(8) yields A+B+C = 0.

We have proved by induction that A+B+C = 0 for any type(i) stream triangle. We leave the similar proof for type(ii) stream triangles as an exercise.

Now let ABCD be a type(i) stream rectangle with cross stream EF as in Figure 7.7(II). We consider the following cases.

(1) *BF is odd and EF is odd.* In this case FC must be even; ED must be odd and AE must be even. Then stream quadrangle EDCF is type(ii) and thus (9) D+C+F = E by the inductive hypothesis. Stream quadrangle AEFB is type(ii) and thus (10) E+A+B = F by the inductive hypothesis. Combining (9), (10) yields A+B+C+D = 0.

(2) *BF is odd and EF is even.* In this case FC is even; ED must be even and AE must be odd. Then stream quadrangle EDCF is type(i), and thus (11) E+D+C+F = 0 by the inductive hypothesis. Stream quadrangle AEFB is type(iii) and thus (12) A+B = E+F by the inductive hypothesis. Combining (11), (12) yields A+B+C+D = 0.

(3) *BF is even and EF is odd.* In this case FC is odd; ED must be even and AE must be odd. Then stream quadrangle EDCF is type(ii), and thus (13) E+D+C = F by the inductive hypothesis. Stream quadrangle AEFB is als type(ii), and thus (14) A+B+F = E. Combining (13),(14) yields A+B+C+D = 0.

(4) *BF is even and EF is even.* In this case stream FC is odd; ED must be odd and AE must be even. Then stream quadrangle EDCF is type(iii), and thus (15)E+F = D+C by the inductive hypothesis. Stream quadrangle AEFB is type(i), and thus (16) A+E+F+B = 0. Combining (15),(16) yields A+B+C+D = 0.

We have proved by induction that for every type(i) stream quadrangle ABCD we have A+B+C+D = 0. Similar methods are used to prove the Theorem stipulations for the three remaining types of stream quadrangles, and we leave these also as exercises. (Beware: odd + odd = odd and even + even = odd (!) in computing stream lengths along the same line.) □

The equations listed in Theorem 7.1 will be referred to as *corner equations.*

We have affixed a (*) at the front of this proof because it is incomplete. The reason is that the cross-stream EF in Figure 7.7(II) may not exist. In Figure 7.8 several examples of stream quadrangles without *splitting cross-streams* are illustrated, (A *splitting cross stream* S in a stream quadrangle Q is one in which two of the corner tiles of Q are on one side and the other two are on the other side of S.)

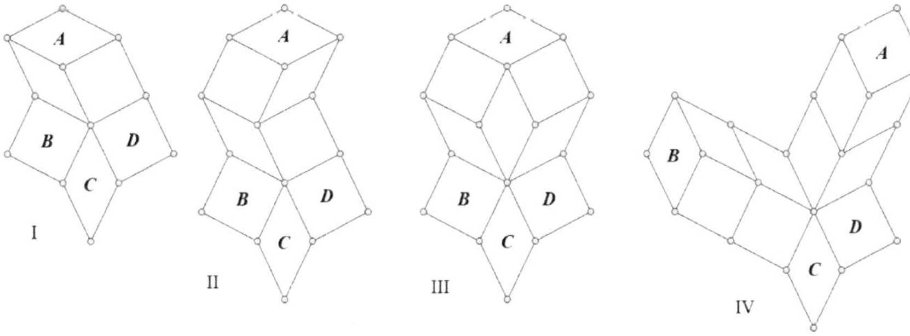

Figure 7.8

The quadrangles in Figure 7.8 admit of various extensions but are all similar in structure: they all have a central pentagon or hexagon. It is not clear what other types of quadrangles exist which are without a splitting cross-stream. To complete the proof of Theorem 7.1, all such stream quadrangles must be determined and shown to satisfy the stipulations of the Theorem. The notation [a,b,c] for a stream triangle or [a,b,c,d] for a stream quadrangle indicates the sequence of stream lengths for the bounding streams. The [2,2,3,3] quadrangle in 7.8(I) satisfies A = B+C+D; the [2,2,4,4] in Figures 7.8(II,III) each satisfy A+B+C+D = 0, and the [2,4,6,4] in 7.8(IV) also satisfies A+B+C+D = 0. These results are all easily seen and agree with the stipulations of Theorem 7.1.

7.3 Free Generating Sets

Based on the results of Section 7.2 it is likely that the corner values of any stream polygon satisfy a corner equation. Thus no FGS for a semiregular polygon P tiled with rhombs can contain all corners belonging to any stream subpolygon. If S is a subset of the tile set of P we define the graph Γ(S) in a fashion similar to the graph Γ(S) defined in Section 6.2 in case P is a rectangular polysquare: the vertices of Γ(S) correspond to the tiles in S, and an edge is inserted between two vertices if they are nearest neighbors in the same tile stream. To avoid containing all corners of any stream subpolygon it is required that Γ(S) be acyclic.

In Figure 7.9 an FGS S is illustrated along with its graph Γ(S), in an octagon P with semiperimeter s = 12. Note that Γ(S) is a tree with exactly s-1 = 11 vertices. It is essential that Γ(S) contain at least one vertex in each tile stream of P, otherwise values in the vacant tile stream could not be generated.

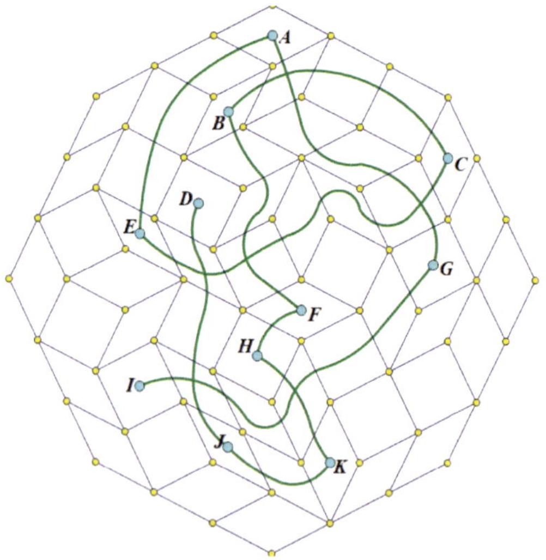

Figure 7.9

If the semiperimeter of P is s, then S must contain exactly s-1 tiles, consistent with Theorem 4.5. we thus have the following result.

Theorem 7.2 *Let P be a semiregular polygon tiled with rhombs and with semiperimeter s, and let S be a subset of the tile set of P. Then S is an FGS for P iff Γ(S) is a tree with exactly s -1 vertices.*

Proof. Analogous to the L-inversions used in Theorem 6.2 to determine FGS's in rectangular polysquares, we define a *deleted cycle inversion.* Suppose Γ(S) is a tree with s-1 vertices. Then Γ(S) has no cycles, but it will have *deleted cycles,* i.e., tree elements C which fall short of forming a cycle by just one missing vertex. A *deleted cycle inversion* on C is achieved by removing any one of the elements of C and adding an element in the place occupied by the missing vertex. If S′ is the resulting set of vertices, then as in theorem 6.2, S will be an FGS iff S′ is. The proof will be completed by finding a sequence of deleted cycle inversions which produce a set S′ which is a simple generating set for P.

A *corner* of the tree Γ(S) is a tile of S which lies on two streams. A corner and two adjacent tiles of Γ(S) form a deleted cycle and thus allow for an inversion. As in the proof of Theorem 6.3, a sequence of deleted cycle inversions can be found which results in a set S′ of tiles which is easily seen to be a simple generating set for P. As a consequence S is also an FGS for P.

Conversely, suppose S is an FGS in P. If Γ(S) contains a cycle, then the corner values of this cycle will satisfy a corner equation so their values cannot be freely chosen and S would not be an FGS. The argument to show that Γ(S) is connected is exactly the same as the argument given before the statement of Theorem 6.4, except that "rows and columns" is replaced by "streams". Thus Γ(S) is a tree, and finally Γ(S) must have exactly s-1 vertices in accordance with Theorem 4.5. □

In Figure 7.10 a sequence of deleted cycle inversions is applied to the FGS in Figure 7.9. In the first step tile H is replaced by tile H′ in the quadrangle [FHKH′]. Also J is replaced by J′ in the quadrangle [DJKJ′]. In the second step C is replaced by C′ in the triangle [CBC′] and D is replaced by D′ in the triangle [DD′J′].

In the third step E is replaced by E′ in the triangle [EC′E′] and G is replaced by G′ in the quadrangle [E′AGG′]. In the fourth step x is replaced by I′ in the quadrangle [IG′Fx] and I is removed. In the final steps G′, H′, K are replaced by G″, H″ and K′ respectively via a few more deleted cycle inversions which the reader can easily find.

The end result is the tile path in Figure 7.10(V) which joins two opposite corner tiles of P. Such a path was shown to be a simple FGS in THeorem 4.4.

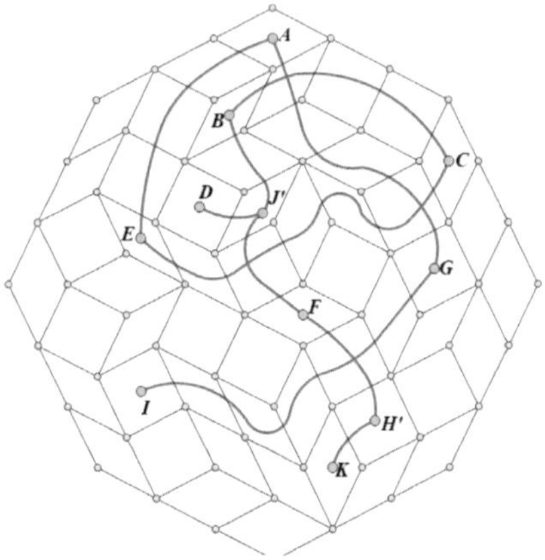

Figure 7.10(I)

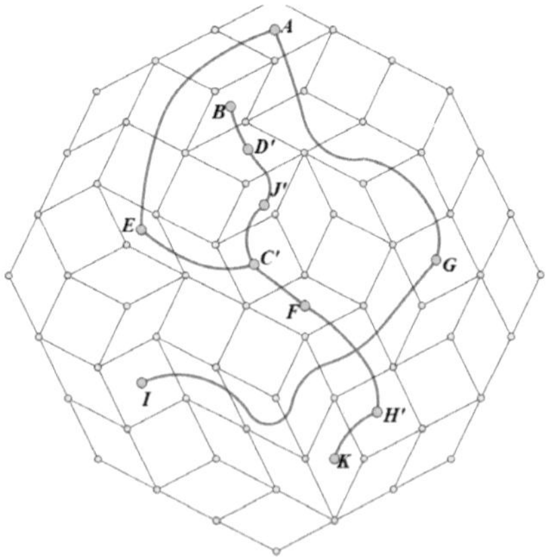

Figure 7.10(II)

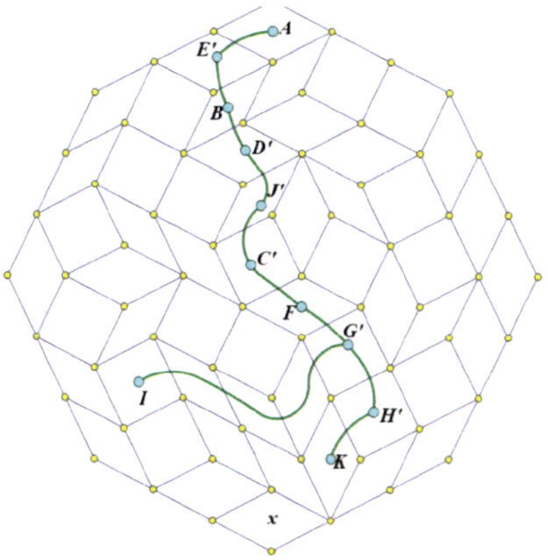

Figure 7.10(III)

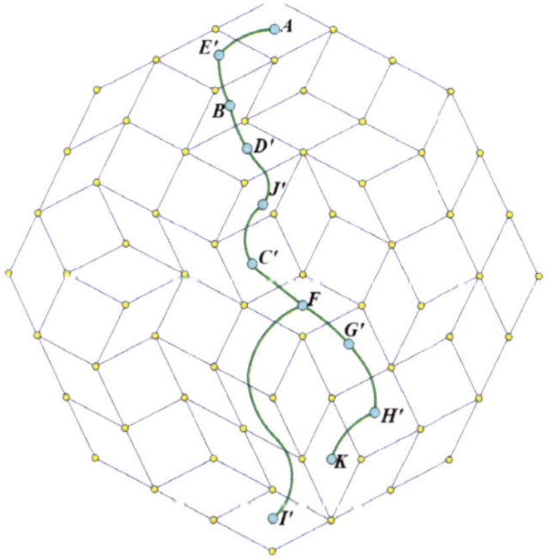

Figure 7.10(IV)

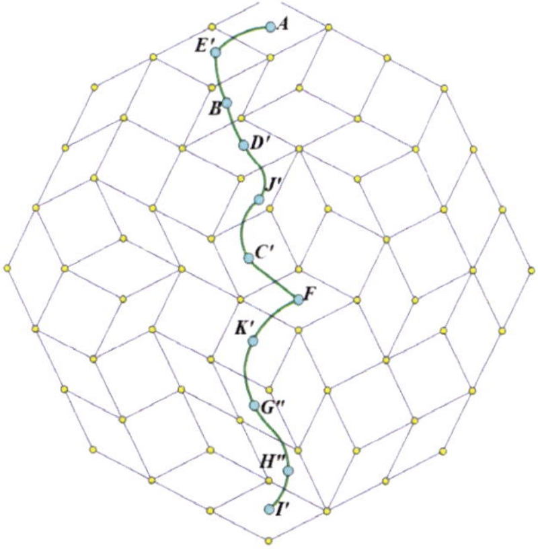

Figure 7.10(V)

7.4 Exercises

1. Show that the stream polygons in Figures 7.5, 7.6 and 7.8 satisfy the stipulations of Theorem 7.1.

2. Stream pentagons are of the types indicated in Figure 7.11.

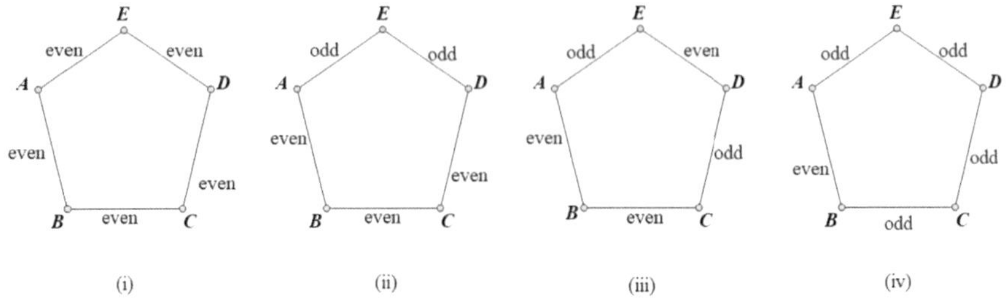

Figure 7.11

For these stream pentagons a *splitting cross-stream* separates the corners two on one side and three on the other. Prove the following Theorem inductively, assuming that all stream pentagons with no splitting cross stream can be shown to satisfy the Theorem stipulations directly.

Theorem 7.2 *For any type(i) stream pentagon A+B+C+D+E = 0; for any type(ii) stream pentagon A+B+C+D = E; for type(iii) stream pentagons A+B+C = D+E; for type(iv) stream pentagons A+B+D = E+C.*

3. In Figure 7.10 several stream pentagons with no splitting cross-stream are illustrated.

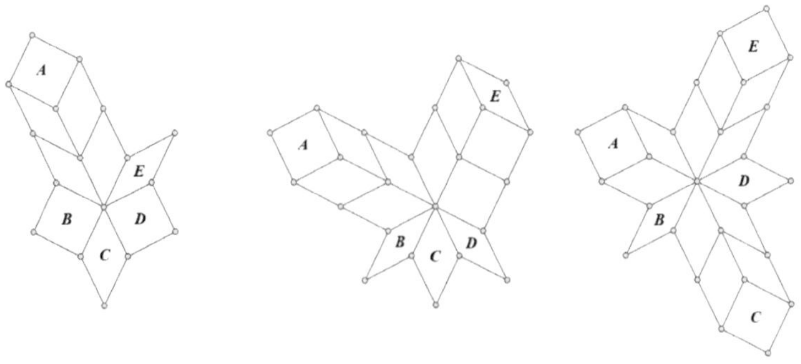

Figure 7.12

The pentagons in Figure 7.12 are similar in that they all feature extensions of a central nbd with 6, 7 or 8 tiles. The extensions do not *communicate* with each other in that there is no cross-stream connecting them. Show that the stream polygons in Figure 7.12 satisfy the stipulations in Theorem 7.2. Does there exist a stream pentagon without cross-streams which is essentially different from the ones illustrated in Figure 7.12?

4. Find and then prove a Theorem similar to Theorems 7.1 and 7.2 which describes corner equations for a stream hexagon. Construct some examples of stream hexagons which have no splitting cross-stream. For a stream hexagon there are two types of splitting cross streams which can be used for an inductive argument, namely those which divide the corners with 2 on one side and 4 on the other, and those which divide the corners equally with 3 on each side. There are 8 types of stream hexagons. Construct a semiregular polygon P tiled with rhombs or parallelograms and provide P with a locally nilpotent assignment. Locate each of the 8 types of stream hexagons in P and use these to conjecture the associated corner equations. (This is how Theorems 7.1 and 7.2 were determined.)

5. Choose integer values freely for tiles A, B, ..., K in Figure 7.9 and use Theorem 7.1 to complete a locally nilpotent assignment for P.

CHAPTER 8

Quadrilateral Tilings

8.1 Q1 Tiles

The five nonparallegram lattice quadrilaterals in Figure 8.1 constitute the set of $\mathbf{Q}_1$ tiles. These consist of all convex nonparallelogram quadrilaterals with edges of length 2, $\sqrt{5}$ or $\sqrt{10}$. These were introduced in [2] where they were used to tile a number of semiregular polygons. Two examples are illustrated in Figures 8.2, 8.3.

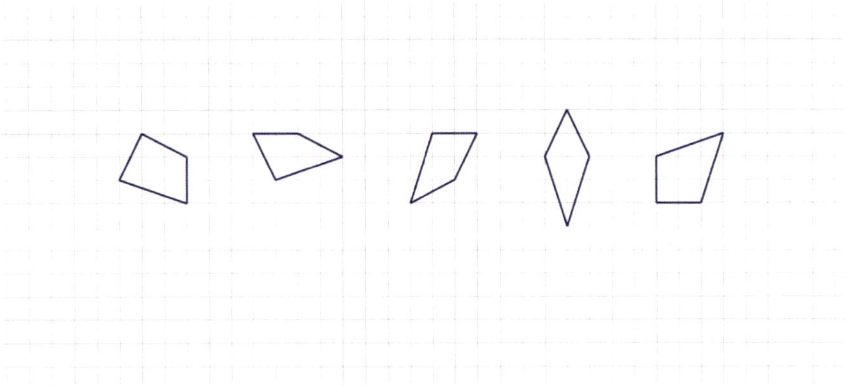

Figure 8.1

Streams can still be defined for $\mathbf{Q}_1$ tilings: instead of passing through a series of mutually parallel sides, the stream just moves to the opposite side of each tile it enters. Stream polygons can thus be defined and the corner equations of Theorem 7.1 remain valid.

If the tiling in Figure 8.2 is given a locally nilpotent assignment, then the corners of the type(ii) stream triangle (A,B,C) satisfy A+C = B. In Figure 8.3 a type(iv) stream rectangle is indicated and the corner equation A+C = B+D will hold in accordance with Theorem 7.1.

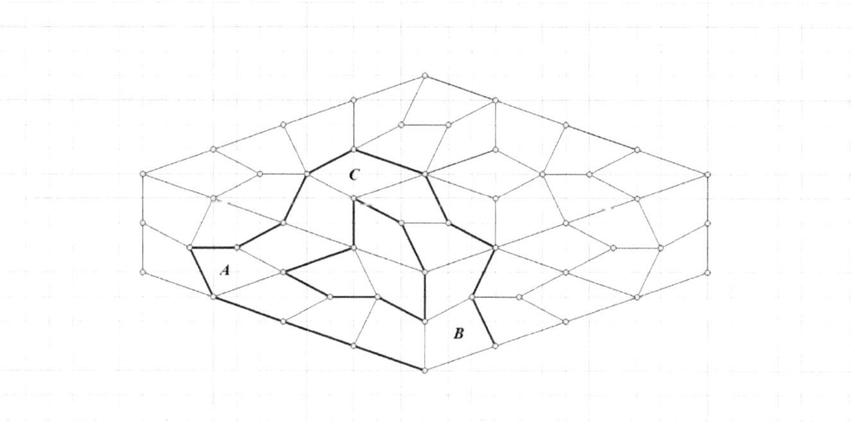

Figure 8.2

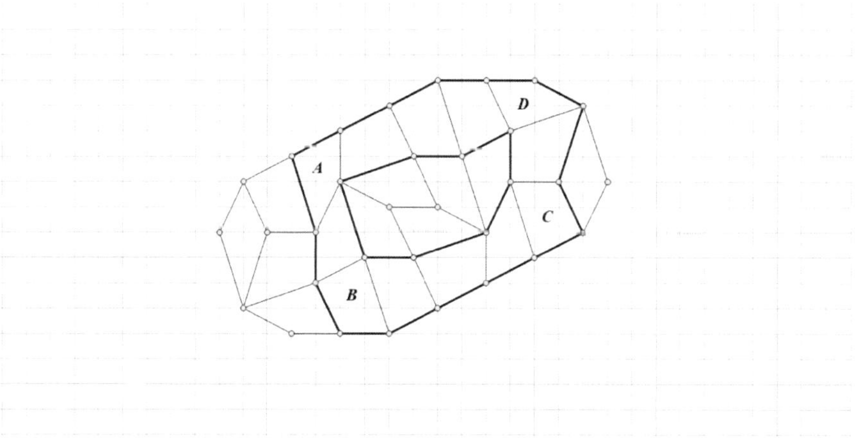

Figure 8.3

The streams in $\mathbf{Q}_1$ tilings also exhibit some properties that cannot be found in parallelogram tilings. For example, two streams can cross each other in more than one tile as in the example in Figure 8.4. The result is a stream polygon bound by two streams, which we call an *eye*.

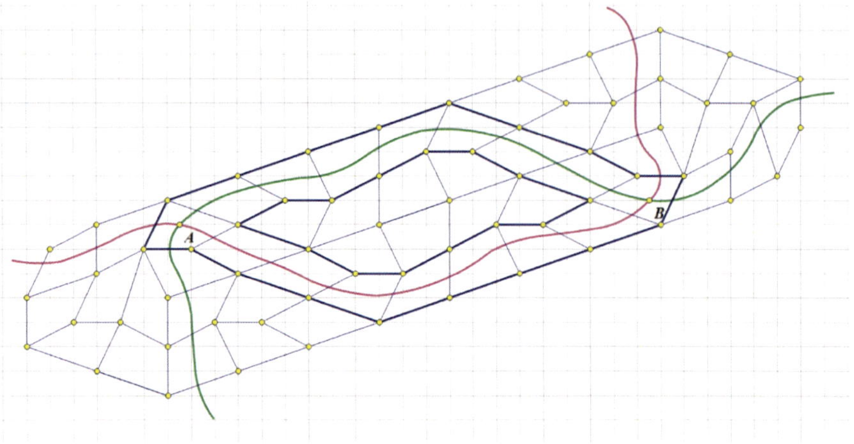

Figure 8.4

Another phenomenon which can occur in $\mathbf{Q}_1$ tilings but not in parallelogram tilings, is the existence of *cyclic streams.* These occur when a single stream bounds a region as in the example in Figure 8.5.

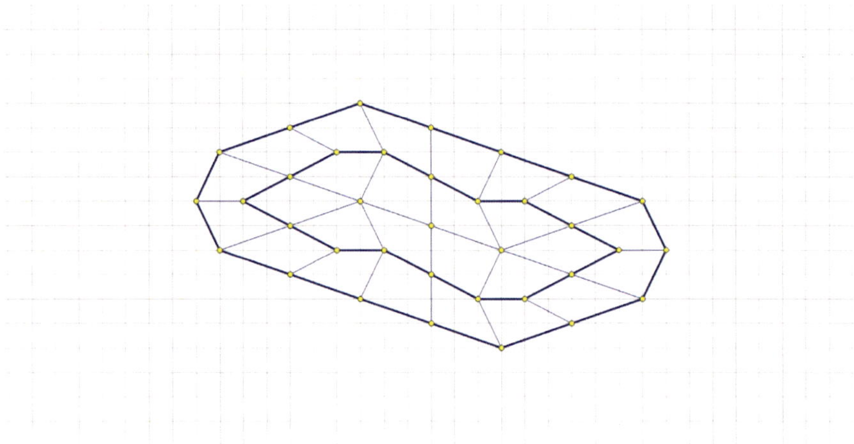

Figure 8.5

8.2 Eyes

The first thing to observe about an eye is that it must consist of an even number of tiles. This is so since every cross stream entering the eye through one edge must depart the eye through another.

If we denote an eye by [A,B] where A, B represent the two tiles where the bounding streams intersect, then the stream segments between A, B are either both odd or both even in length. The following Theorem will be proved inductively in a manner similar to the proof of Theorem 7.1.

Theorem 8.1 *In the eye [A,B], if both stream segments from A to B are even then A+B = 0, and if both stream segments are odd then A = B.*

Proof. The diagrams in Figure 8.6 represent the four possibilities in case both stream segments joining A, B are odd.

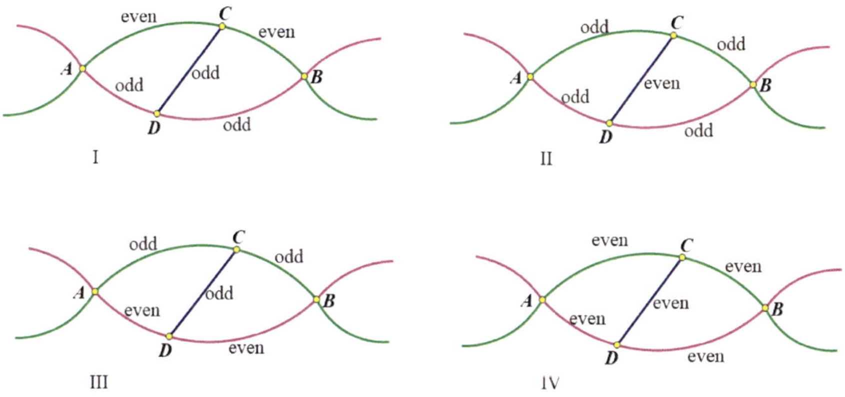

Figure 8.6

In Case(I) we assume stream segment AD is odd and cross-stream CD is odd. Then DB must be odd; AC must be even and CB must be even. Both [ADC] and [CDB] are type(ii) stream triangles, so by Theorem 7.1 we have A+D = C and D+B = C. Combining these equations yields A = B.

In Case(II) we assume AD is odd and CD is even. Then DB must be odd and AC and CB must be odd as well. Then [ACD], [CDB] are both type(ii) stream triangles and thus C+D = A and C+D = B, again yielding A = B.

In Case(III) we assume AD is even and CD is odd. Then BD must be even and AC, CB must both be odd. Then stream triangles [ACD], [CDB] are both type(ii) and thus A+D = C and D+B = C. Combining these equations yields A = B.

In Case (IV) we assume AD and DC are both even. Then AC, CB, DB must all be even. Triangles [ACD], [CDB] are type(i), so by Theorem 7.1 we have A+C+D = 0 and C+D+B = 0. These equations again yield A = B.

A similar argument can be used in case both stream segments joining A, B are even. We will omit the details. □

8.3 Even Stream Polygons

A stream polygon each of whose bounding streams has even length will be called *even*. A tile path ρ is *jagged* in case each constituent tile is a corner. Equivalently, a jagged tile path consists of a series of streams of length 2. A *jagged polygon* is a stream polygon bound by a closed jagged path. In Figures 8.7, 8.8 even stream polygons are illustrated in a $\mathbf{Q}_1$ polygon. The one in 8.8 is jagged. The same locally nilpotent assignment is given in both Figures. Note that the corner values sum to 0 in both cases.

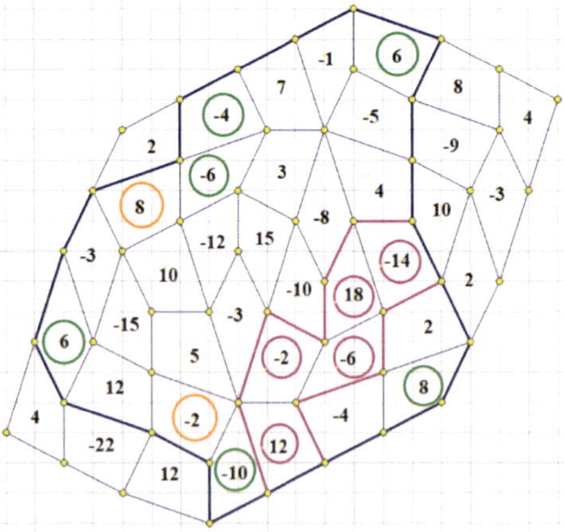

Figure 8.7

In Figure 8.7 a jagged path indicated in pink, is illustrated inside an even stream hexagon. The pink path has the same value as the corner of the hexagon it encloses. The pink path along with the remainder of the stream hexagon forms an even stream polygon with smaller order than the original stream hexagon, but whose corner sum is the same as the original hexagon, namely 0. Sometimes a corner can be cut using just two tiles. In Figure 8.7 the indicated yellow tiles must have the same value as the corner, (with value 6), that they cut. The two yellow tiles along with the remaining, (green), tiles of the hexagon form the corners of an even stream polygon.

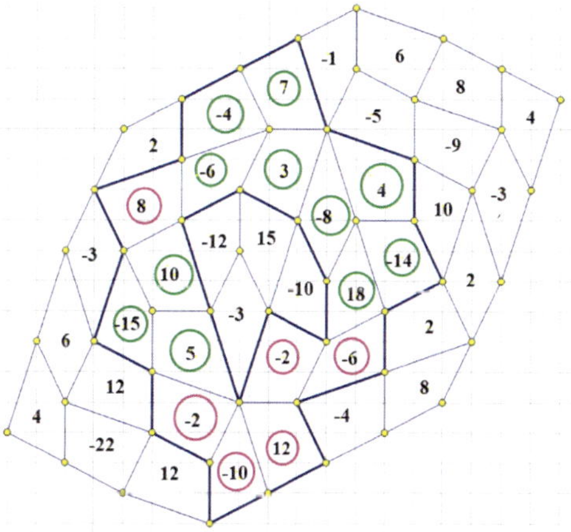

Figure 8.8

It is not possible to cut any corners in the jagged stream polygon P in Figure 8.8, but removal of the nbds leaves the pink tiles which form an even stream polygon with the same sum, (0), as P. These examples suggest an inductive argument which we use to prove the following Theorem.

Theorem 8.2 *The corner sum of every even stream polygon is 0.*

Proof. (*) We proceed by induction on the order of an even stream polygon. As a basis observe that every nbd is itself an even stream polygon and thus its corners, (i.e., all the tiles in the nbd), have 0 sum in a locally nilpotent assignment. If P is an even stream polygon, we hope to proceed by induction by locating within P a smaller even stream polygon with the same corner sum, (0), as P. This may be accomplished in either of the two ways indicated in Figures 8.7. 8.8. That is, either by cutting a corner as in Figure 8.7, or by removing nbds to obtain a smaller even stream polygon as in Figure 8.8.

We again arrive at the difficulty of specifying precisely the full set of even stream polygons for which no such operations can be performed. Consider, for example, the series of cutting operations indicated in Figure 8.9.

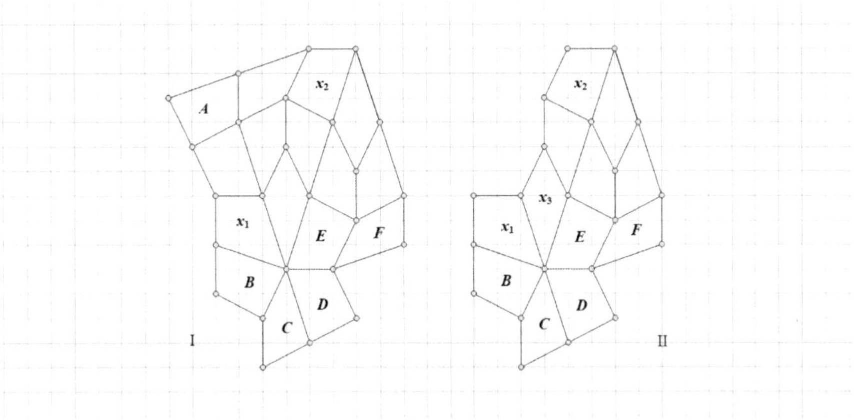

Figure 8.9

In the even stream hexagon [ABCDEF] in Figure 8.9(I) it is easily seen that $A = x_1 + x_2$. Thus the corner A can be cut resulting in the even stream heptagon [x_1BCDEFx_2] in Figure 8.9(II) which will have the same corner sum as [ABCDEF]. The polygon in 8.9(II) cannot be reduced further, so it and others like it should be included in a basis. This polygon, however, is easily seen to have 0 corner sum: stream triangle [x_2x_3F] is type(ii), so by Theorem 7.1, $x_2 + F = x_3$. Then the corner sum for 8.9(II) is $x_2 + F + E + D + C + B + x_1 = x_3 + E + D + C + B + x_1 = 0$. Since we have not determined precisely the configurations which constitute the basis, the proof is incomplete.

Otherwise, the cutting operations allow the induction to be completed. □

8.4 Nilpotent Q1 Polygons

As with polygons tiled with rhombs, $\mathbf{Q}_1$ polygons with a nbd dissection are nilpotent. Such a $\mathbf{Q}_1$ polygon is illustrated in Figure 8.10(I). Unlike polygons tiled with rhombs, hexagon inversion *does not* preserve nilpotence. The polygon in Figure 8.10(II) is obtained from 8.10(I) by the indicated hexagon inversion, and it is not nilpotent due to the existence of an odd length stream. (Theorem 1.7 holds equally well for $\mathbf{Q}_1$ polygons.) The polygon P in Figure 8.10(III) is obtained from 8.10(II) by a further hexagon inversion. All streams are now even and the resulting $\mathbf{Q}_1$ polygon is nilpotent. This can be seen by the indicated dissection into nbds and the corners of the even stream quadrangle [A′EC′D′]. We will argue that such a dissection is always possible in case every stream in P is even.

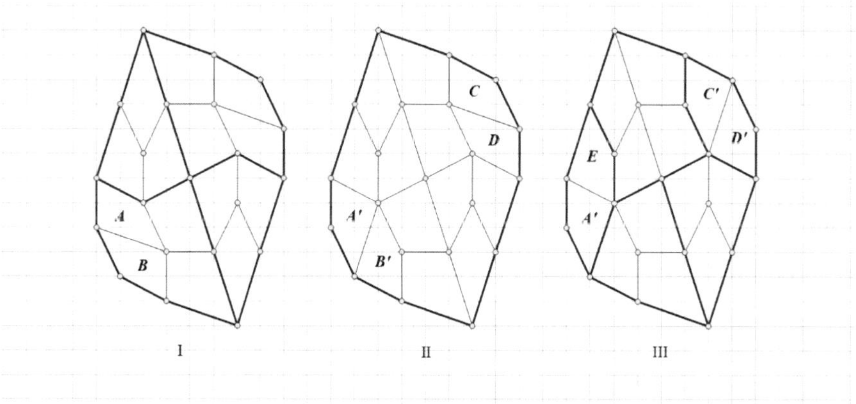

Figure 8.10

Theorem 8.3 *Let P be a semiregular polygon with an edge to edge tiling consisting of* $\boldsymbol{Q}_1$ *tiles and parallelograms. Then P is nilpotent iff every stream in P is even.*

Proof. Assume that every stream in P is even. We will create a dissection of P into nbds and sets of corners of even stream polygons. As a first step it is convenient to find as many nbds as possible and include these in the dissection. These are indicated in bold outline in the example in Figure 8.11. The tiles not contained in these nbds will be called *residual.* In Figure 8.11 the residual tiles are circled. We will argue that the residual tiles form the corners of one or more even stream polygons.

Since every stream in P is even and passage through any number of nbds in the dissection adds an even number of tiles to the stream, it must be that both streams through a residual tile A contain a second residual tile. Thus if S_1, S_2 are the two streams through A, then S_1 must contain a residual tile B, and S_2 must contain a residual tile F. Then both stream segments AB, AF are even. It follows similarly that streams through tiles B, F contain residual tiles, B′, F′ respectively with stream segments BB′, FF′ both even. Continuing in this way an even stream polygon containing corner value A is obtained

If more residual tiles remain, then the argument is repeated to obtain a second even stream polygon. Ultimately a partition of the residual tiles of P into corners of even stream polygons is obtained. By Theorem 8.2 these corners have sum 0 and since the tiles in each nbd in the dissection also have 0-sum in any locally nilpotent assignment for P, we can conclude that P is nilpotent.

Conversely, if P has an odd stream, then by Theorem 1.7, (augmented to include semiregular polygons tiled with convex quadrilaterals), P is not nilpotent. □

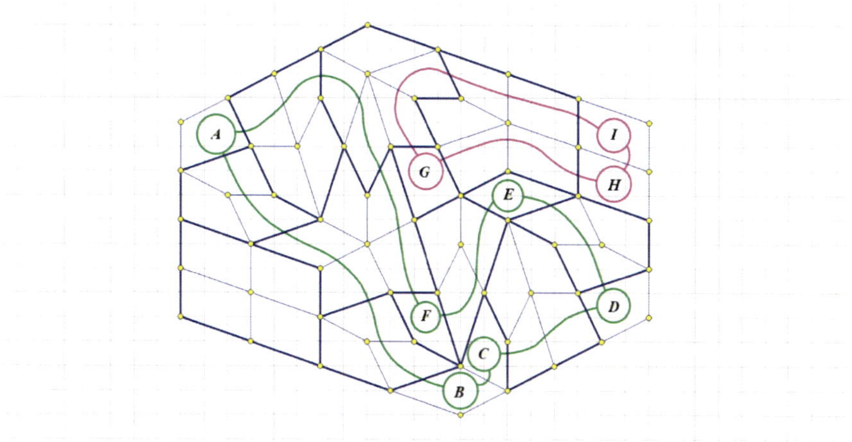

Figure 8.11

In the example in Figure 8.11, the residual tiles form the corners of two even stream polygons, one in green and the other in pink. Note that the semiregular polygon in this Figure was obtained by fusing together two copies of an even $\mathbf{Q}_1$ polygon and adding parallelograms in the corners. There is no need to restrict Theorem 8.3 to $\mathbf{Q}_1$ polygons since it remains equally valid for any edge-edge tiling involving $\mathbf{Q}_1$ tiles and parallelograms. Let $\mathbf{CQ}(2, \sqrt{5}, \sqrt{10})$ denote the set of convex quadrilaterals with sides of length 2 or $\sqrt{5}$ or $\sqrt{10}$. The semiregular polygon in Figure 8.11 is a member of this class.

8.5 Exercises

1. In Figure 8.12, a semiregular octagon P is tiled with rhombs and a partial nbd dissection is indicated. Residual tiles are labeled A, B, ..., I. We know from Theorem 1.6 that P is nilpotent. This was proved using hexagon inversion. Since all streams of P are even, the method of Theorem 8.3 can also be used to prove nilpotence. Find the even stream polygon(s) with corners consisting of the residual tiles. Then provide a locally nilpotent integer assignment to the tiles of P, and check that the values assigned to the corners have 0 sum.

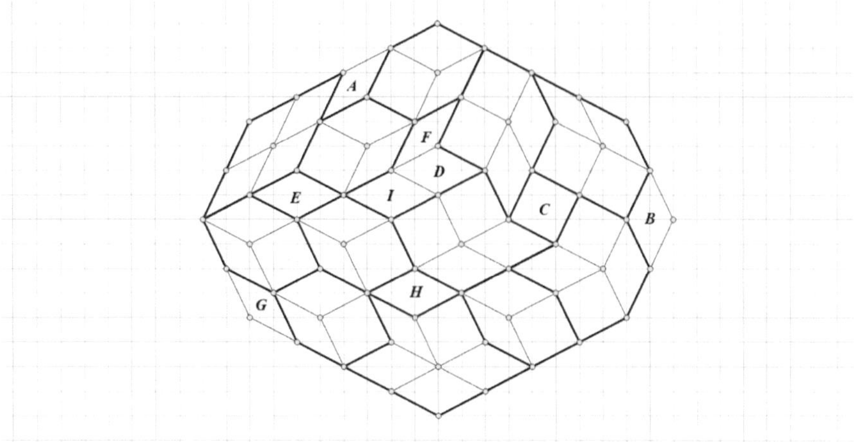

Figure 8.12

2. In Figure 8.13 a $\mathbf{Q_1}$ tiling of a semiregular polygon P with semiperimeter s = 11 is illustrated. For the subset S = {A, B, … , J} of tiles the associated graph Γ(S) is drawn. Γ(S) is a tree with s-1 = 10 vertices. Assign integers freely to the tiles in S and show that your assignment can be extended to an FGS for P. Theorem 7.2 can be generalized to include semiregular polygons tiled with convex quadrilaterals.

3. In Figure 8.14, a semiregular octagon P is tiled with rhombs. The ten corner tiles of a stream decagon D are labeled $A_1B_1A_2B_2A_3B_3A_4B_4A_5B_5$. Each bounding stream of D has odd length. Provide a locally nilpotent integer assignment to P and check that $A_1+A_2+A_3+A_4+A_5 = B_1+B_2+B_3+B_4+B_5$. If $S = A_1B_1 \ldots A_nB_n$ is any stream polygon with an even number of bounding streams each of which is odd, show that $\Sigma\{A_i : i = 1, \ldots ,n\} = \Sigma\{B_i : i = 1, \ldots ,n\}$ for any locally nilpotent assignment.

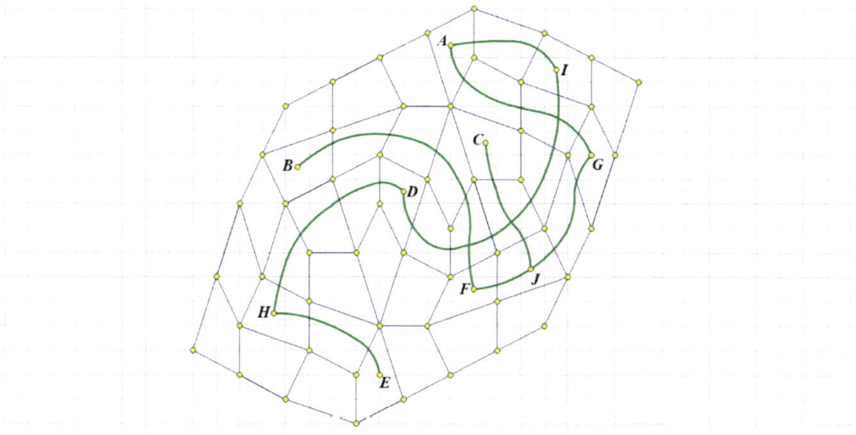

Figure 8.13

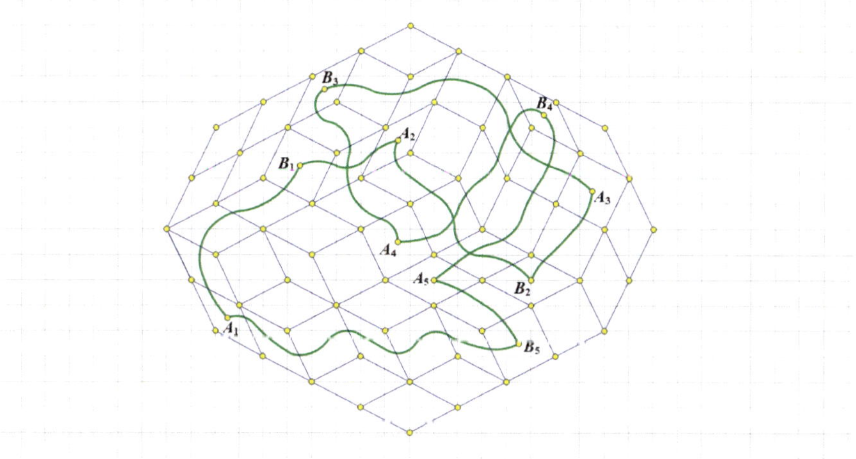

Figure 8.14

4. Prove the equation in problem (3) for a stream polygon with all odd sides in a rectangular polysquare. As an example, consider the stream polygon $A_1B_1A_2B_2A_3B_3A_4B_4A_5B_5$ in Figure 8.15. For this stream polygon prove inductively that $\Sigma\{A_i : i = 1,2,3,4,5\} = \Sigma\{B_i : i = 1,2,3,4,5\}$ for any locally nilpotent assignment.

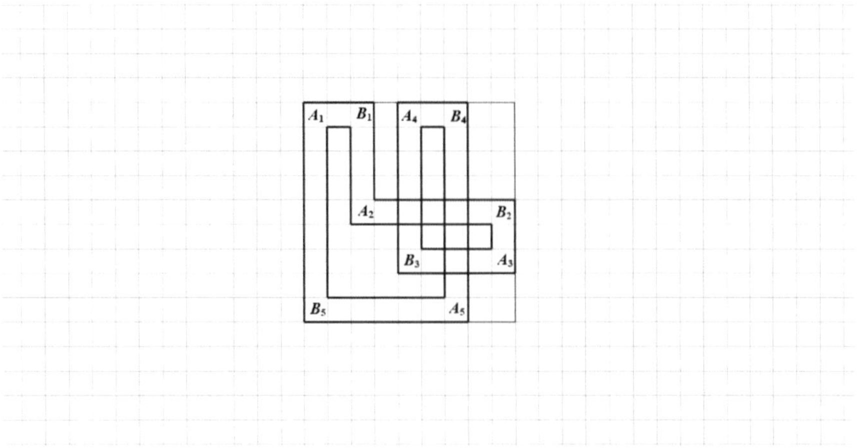

Figure 8.14

5. Can Theorem 8.3 be strengthened as follows?

Conjecture 8.4 *Let P be any polygon with an edge to edge tiling with convex quadrilaterals. Then P is nilpotent iff every stream in P is even.*

The nonsemiregular polyrhomb in Figure 8.15 has the property that every stream has even length. Show that this polyrhomb is nilpotent.

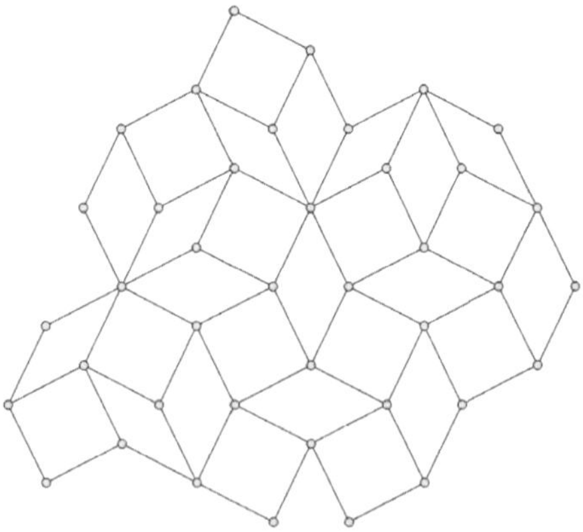

Figure 8.15

CHAPTER 9

The Game of Rhombs

Rhombs is a game similar to dominoes but is played with tiles in the shape of rhombs. These rhombs have interior angles of 60° and 120° and can be used to form hexagons and stars as in Figure 9.1. In the *9-version* each rhomb is marked with dots representing the numbers 1, ... , 9. Blank tiles are also included with a value of 0. There are 6 rhombs of each denomination including 6 blanks for a total of 60 tiles. The object of the game is to make scores by completing a hexagon or star with rhombs whose values sum to a multiple of 5. For example, the simple hexagon in Figure 9.1 yields a score of 10, and the star in Figure 9.1 scores 30.

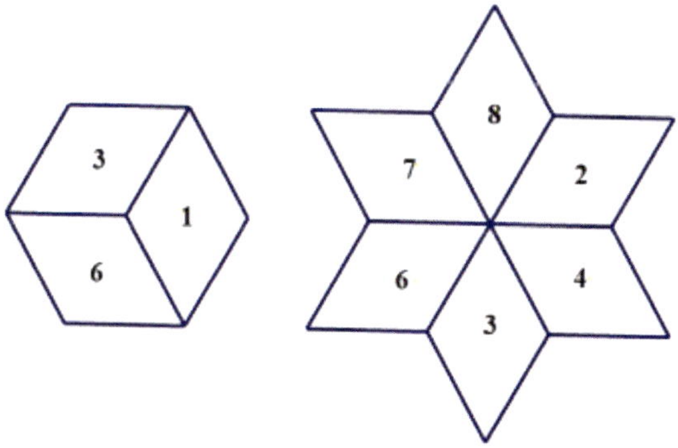

Figure 9.1

Suppose there are two players: player A and player B who take alternate turns. To start the game, all rhombs are placed face down in the *boneyard*. Both players select 8 tiles from the boneyard to form their *hands* and set them upright to view them. Player A goes first by selecting a tile from his/her hand and placing it face up at the table center. No score is possible, so play passes to player B. B plays one of his/her rhombs adjacent to the played tile so as to form a *chevron* as in Figure 9.2(I). Still, no score is possible. *It is never allowed to place a rhomb which forms a parallelogram, thus the play indicated in Figure 9.2(II) is forbidden.*

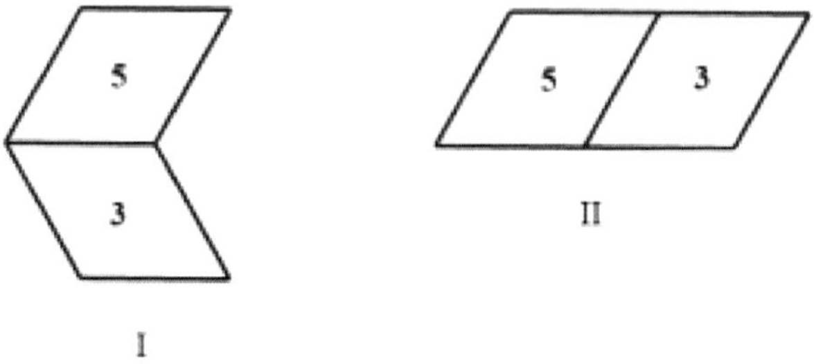

Figure 9.2

Before each turn players draw from the boneyard to replenish their hands to a full complement of 8 tiles. At his/her second turn, A has an opportunity to make a score by completing a hexagon with total value a multiple of 5. Suppose the first two played tiles are as in Figure 9.2(I). If now A has a tile with value 7 in hand, then it can be played to form a hexagon as in Figure 9.3(I) yielding a score of 15, or if A has a tile with value 2 then it can be played to form a hexagon with a score of 10. *No hexagon or star can be completed which does not yield a score.* Thus in Figure 9.3(I) only a 7-tile or a 2-tile can be played to form a hexagon; no other numbered tile can be put in that position.

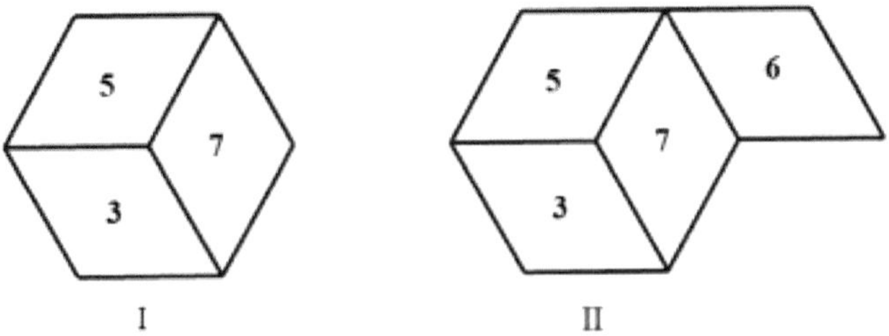

Figure 9.3

Whenever a player makes a score then he/she can make another play *in the same turn.* Thus it is possible to play several scoring tiles in the same turn. This is done *without replenishing the hand.* When no more scores can be made, (or the player chooses not to make a score), then one more nonscoring tile must be played before the turn is completed. Thus A must play one additional (nonscoring) tile as in Figure 9.3(II). It is possible that all 8 tiles in a players hand can be played in one turn, in which case the turn is also completed possibly without playing a nonscoring tile. Only when a turn is completed is a player allowed to replenish his/her hand by drawing from the boneyard.

A *set* is completed when all 60 tiles have been played. If the boneyard is empty and a player has no tiles in hand, then she/he must *pass,* and the other player can play any remaining tiles in her/his hand. Scores are then tallied and the player with the highest score is the winner of the set. Any number of sets can be played with A, B alternately making the first play. The player with the largest combined score from all sets is the winner.

As an example of a possible turn, suppose the game configuration (or *polyrhomb*) appears as in Figure 9.4(I) and it is player B's turn. Suppose also that B has tiles with values {1, 2, 3, 4, 5} in hand. These tiles can be placed by B in the sequence (2, 3, 5, 4, 1) to obtain the polyrhomb in Figure 9.4(II). Placing first the 2- tile yields a score of 10, and B earns an extra play. Then B places the 3-tile to create a star with score of 25, and thus earns another extra play. B then places the 5- tile and 4-tile in succession to obtain an additional 20 points. B then completes his/her turn by playing the nonscoring 1- tile. This tile can be played to form a chevron with any boundary tile. B's total score is 55 for this turn. Since B has played 5 tiles, B selects 5 tiles from the boneyard so that his/her hand is restored to a total of 8 tiles. Play then passes to A who has only one possibility for making a score by playing a 3 or an 8 and completing a (1,1,3) or (1,1,8) hexagon.

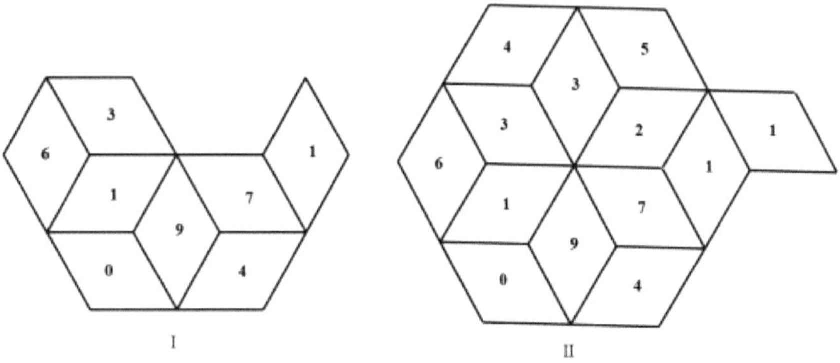

Figure 9.4

By playing a single tile it is possible to complete more than one scoring configuration. In Figure 9.5(I) a tile with value 1 or 6 may be placed in the position marked X thus completing both a star and a hexagon. If X has value 1, then the star has value 25 and the hexagon has value 10 yielding a total score of 35. If X is covered with a 6 then the star gives a score of 30 and the hexagon a score of 15 yielding a total score of 45 for the play. Figure 9.5(II) is exactly like 9.5(I) except that the top tile has a value of 3 instead of 2. For this configuration *no* tile can be placed at position X since it is impossible to complete both the hexagon and the star so that each has a value which is a multiple of 5.

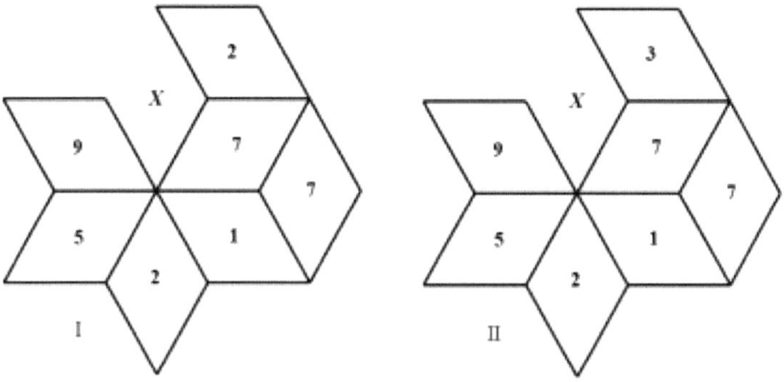

Figure 9.5

There are several more scoring configurations any or all of which players may choose to allow. One consists of 12 tiles and has the shape of a regular hexagon. We call it a *star- hexagon* since it is an extension of a star by means of 6 boundary rhombs. In Figure 9.6(II) one such star hexagon is illustrated. To be a game configuration each constituent simple hexagon, of which there are 6, as well as the enclosed star must each have values which are multiples of 5.

Suppose it is A's turn and A has a 6-tile in hand. Then A can play it in the position marked X in Figure 6(I). *It is only necessary that the 6-tile and the two adjacent tiles, (a 5-tile and a 9-tile in our example), make a simple hexagon with value a multiple of 5.* It can then be shown that the value of the entire created star-hexagon is also a multiple of 5, and thus makes a score. The score consists of the sum of all 12 tiles which constitute the star-hexagon. In our example A will obtain a score of 55 by playing the 6-tile in position X.

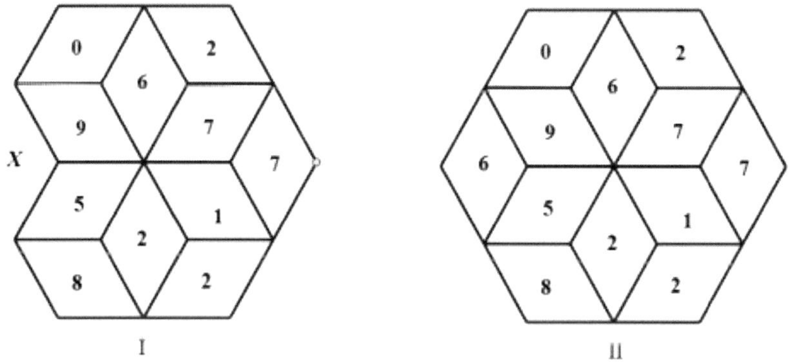

Figure 9.6

In Figure 9.7 the construction called a *stream* is illustrated. A stream is a sequence of adjacent rhombs all of whose shared sides are parallel. The *length* of a stream is the number of tiles it contains. Thus in Figure 9.7, stream A has length 6 and stream B has length 5. The Figure also contains streams of lengths 2, 4 which are not marked. In the evolving game configuration, a player is *not* allowed to place a tile which creates a stream with length greater than 10. The reason for this final rule is that it forces the game configuration to be more compact, thus allowing for the emergence of scoring structures larger than the simple hexagon.

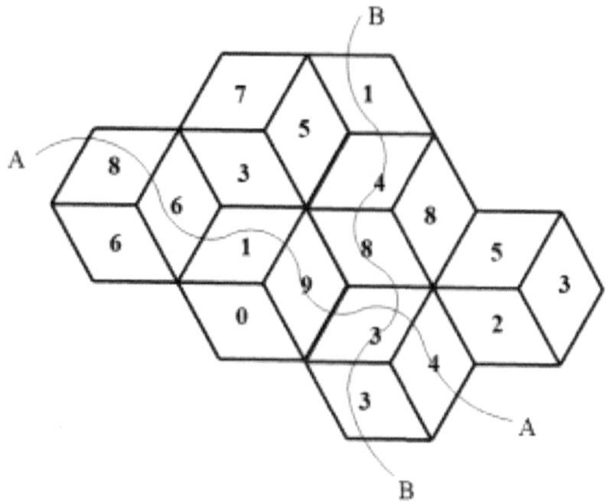

Figure 9.7

An *11-version* with tiles having values 0,1, ... , 11 also makes a good game. Instead of 6 tiles of each denomination, there are only 5. Consequently the number of tiles is 60, the same as in the 9-version. The rules are exactly the same as in the 9-version. Either version can be played with 3 or 4 players who take turns in cyclic fashion. For 3 or 4 players hands should consist of 6 instead of 8 tiles. On average, the time required for a single set in either version is 30 minutes; (longer if players require more thinking time.)

Another possibility is a *14-version* with 4 each of tiles with values 0, ... , 14. Again the total number of tiles is 60 and the rules are the same as in the 9 or 11-version.

CHAPTER 10

Nilpotent Polytriangles

10.1 Introduction

By *polytriangle* we mean a figure composed of unit equilateral triangles fitted together edge to edge. These are called *polyiamonds* in [5] where monohedral tilings of the plane by polyiamonds are discussed. Our concern here will be with nilpotent properties of various polytriangles. In this Chapter we begin by considering polytriangles whose boundary is a triangle, parallelogram, trapezoid or hexagon. For each such figure F, a *locally nilpotent* assignment will be made. By *locally nilpotent* we mean not only that the sum of the six unit triangles surrounding every internal vertex of F is 0, but also that each triangle with two units per side has sum 0. The hexagon consisting of the six unit triangles will be denoted by H_1 and the aforementioned triangle will be denoted by Δ_2. In Figure 10.1 two examples of triangular polytriangles are illustrated, each with a locally nilpotent assignment from the group Z of integers.

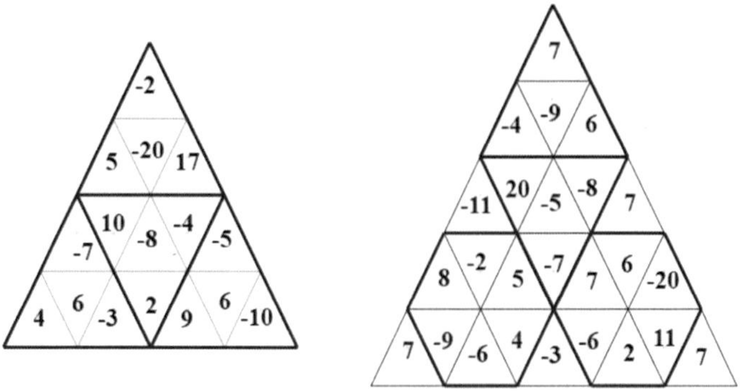

Figure 10.1

We will use the notation Δ_m to denote a triangular polytriangle with m unit edges per side. The Δ_4 at left of Figure 10.1 has a Δ_2 dissection. By definition every Δ_2 is nilpotent, and thus the Δ_4 is nilpotent. The Δ_5 at right in Figure 10.1 is not nilpotent as can be seen by the partial $\{\Delta_2, H_1\}$ dissection which leaves 5 residual tiles with nonzero sum.

10.2 Nilpotent Triangular Polytriangles

As can be surmized from Figure 10.1, Δ_m will be nilpotent if m is even since there will exist a Δ_2 dissection. If m is odd then Δ_m is nonnilpotent as we show in the following Theorem.

Theorem 10.1 *The triangular polytriangle Δ_m is nilpotent iff m is even.*

Proof. If m is even, then there exists a Δ_2 dissection and thus Δ_m is nilpotent. If m is odd consider the locally nilpotent assignment given in Figure 10.2(I). This type of assignment will be called a *linear pattern over {0,1,-1}.*

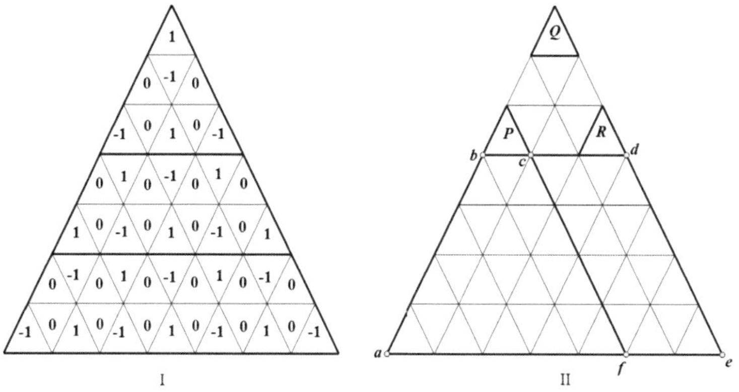

Figure 10.2

In Figure 10.2(I) bold lines indicate the boundaries of Δ_3, Δ_5 and Δ_7. If P is a polytriangle with a locally nilpotent assignment from Z, we will use the notation ν(P) to denote the *value*, i.e., the sum of the values assigned to the unit triangles which constitute P. Thus in Figure 10.2(I), $\nu(\Delta_3) = -1$; $\nu(\Delta_5) = 1$, and $\nu(\Delta_7) = -1$. If this series is continued, the values of the odd base triangles alternate between 1 and -1. We can therefore conclude that Δ_m is nonnilpotent for every odd integer m. □

In the Δ_7 of Figure 10.2(II) the isosceles trapezoids (abcf) and (abde) will be shown in the next Section, to have value 0. We will denote these by T(1,4,5) and T(3,4,5) respectively where the parameters 1,4,5 and 3,4,5 represent the side lengths with the first and last parameters representing the bases and the middle parameter the common length of the lateral sides. Thus ν(T(1,4,5) = ν(T(3,4,5) = 0, and as a consequence the formulas $\nu(\Delta_5)$ = P and $\nu(\Delta_7)$ = P+Q+R can be quickly determined.

10.3 Nilpotent Polytriangular Trapezoids

We will use the notation P(a,b) to denote a parallelogram with sides of length a, b which is tiled with equilateral triangles. In case a and b are both even than P(a,b) has a Δ_2 dissection and consequently is nilpotent. Also if a, b, c are all even then the isosceles trapezoid T(a,b,c) has a Δ_2 dissection and thus is nilpotent. With the next few results it will be shown that some nilpotent trapezoids with odd side lengths also exist.

Lemma 10.2 *In the isosceles trapezoid T(1,2,3) in Figure 10.3(I), X = N + C + P.*

Proof. Consider:

(1) N = -(L+J+C) = X+M+K, and
(2) P = -(M+K+C) = X+L+J, so
(3) N+P+C = (X+M+K) + (X+L+J) + C = (X+X) + (M+K+L+J+C) = (X+X) - X = X. □

Lemma 10.3 *In Figure 10.3(II), the value of triangle [ABC] is X.*

Proof. First note that in the P(2,2) parallelogram with end triangles A, R, it must be that R = -A, and similarly in the P(2,2) parallelogram with end triangles B, Q, we have Q = -B. Then ν([ABC]) = A+B+C = (C+D+F+K+P) + (N+J+C+D+E) - (J+K+D) =(C+D+F) + P + N + C + E = (-E) +P+N+C+E = P+N+C. By Lemma 10.2 P+N+C = X, so we have ν([ABC]) = X. □

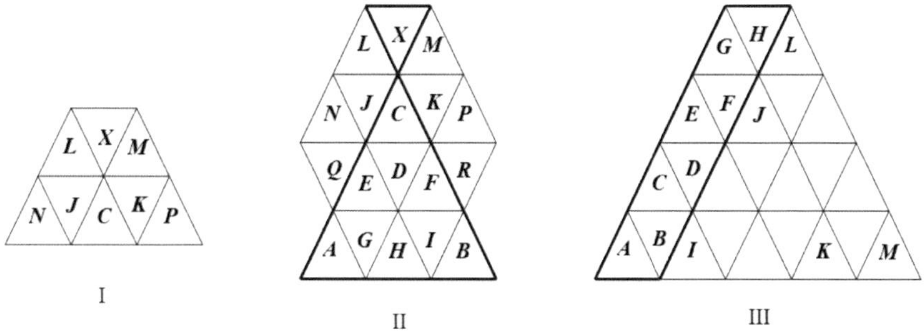

Figure 10.3

Theorem 10.4 *The T(1,4,5) trapezoid is nilpotent.*

Proof. In Figure 10.3(III), ν([GAK]) = 0 since this triangle has a Δ_2 dissection. Then

0 = ν([GAK]) = A+B+C+D+E+F+G + ν([IJK]),

and ν([IJK]) = H by Lemma 10.3. Thus A+B+C+D+E+F+G+H = 0, so the indicated P(1,4) parallelogram has value 0. The T(1,4,5) trapezoid in Figure 10.3(III) has value 0 since it has a dissection into a P(1,4) parallelogram with value 0 and the Δ_4 polytriangle [LIM]. □

Note that a P(1,4) parallelogram is certainly not nilpotent when considered by itself. We have shown that it has 0 value only when considered as a section of a T(1,4,5) trapezoid. In this case we will say that the P(1,4) is *relatively nilpotent.* A polytriangle which is nilpotent when considered by itself is *intrinsically nilpotent.* By using Δ_2 triangles and additional T(1,4,5) trapezoids the following Corollary is immediately obtained.

Corollary 10.5 *Every isosceles trapezoid with lateral sides a multiple of 4 is nilpotent.* □

In Figure 10.4 a construction of the nilpotent T(5,8,13) trapezoid is illustrated.

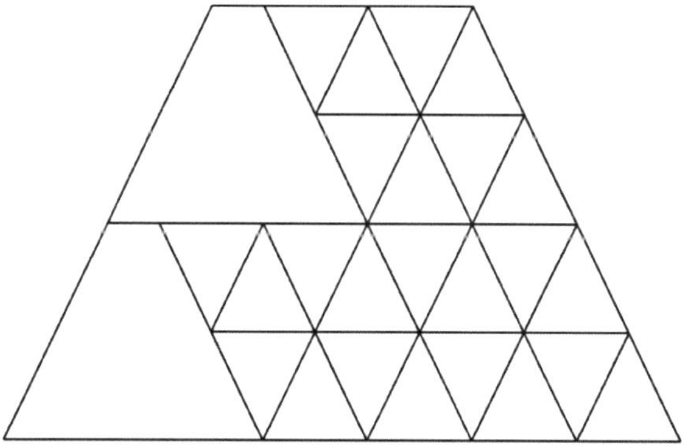

Figure 10.4

Theorem 10.6 *A triangulated isosceles trapezoid is nilpotent iff its lateral sides have length a multiple of 4.*

Proof. It remains only to show that a trapezoid with lateral sides not a multiple of 4 is nonnilpotent. In Figure 10.5 the linear pattern is again used to obtain a locally nilpotent assignment.

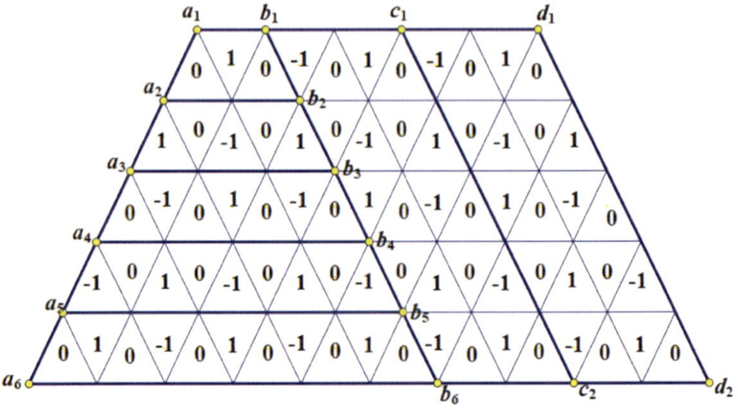

Figure 10.5

The trapezoids $(a_1b_1a_2b_2)$, $(a_1b_1a_3b_3)$, $(a_1b_1a_4b_4)$, $(a_1b_1a_5b_5)$, $(a_1b_1a_6b_6)$, ... in Figure 10.5 have values in sequence which follow the pattern {1,2,1,0,1,2,1,0,1,2,1,0, ...}. The zeros in this sequence occur only in trapezoids with lateral side lengths a multiple of 4. Trapezoids with upper base any odd number can be obtained by the annexation of P(2,n) parallelograms like the parallelograms $(b_1c_1b_6c_2)$ and $(c_1d_1c_2d_2)$ in Figure 10.5. These appended parallelograms all have value 0 in our linear pattern. As a consequence all isosceles trapezoids with lateral sides not a multiple of 4 are nonnilpotent. □

10.4 Nilpotent Triangulated Parallelograms

In case a, b are even, the parallelogram P(a,b) has a Δ_2 dissection, and thus is nilpotent. There also exist nilpotent parallelograms in which one of the dimensions is odd.

Theorem 10.7 *The triangulated parallelogram P(3,4) is nilpotent.*

Proof. With reference to Figure 10.6, the indicated partial dissection into two nilpotent P(2,2) parallelograms shows that P(3,4) will be nilpotent if it can be shown that P+Q+R+S+T+U+V+W = 0. note that

(1) $P+Q+R = -(x_1+x_4+x_5) = y_2$ and
(2) $U+V+W = -(x_2+x_3+x_6) = y_1$.

Then from (1), (2) we have

$P+Q+R+S+T+U+V+W = y_2 + S+T+ y_1 = (y_2+T) + (y_1+S)=$
$-(x_1+x_2+x_4+U) - (x_1+x_2+x_3+R) = -(x_1+x_2+x_4) - U -(x_1+x_2+x_3) - R =$
$R - U + U - R = 0.$ □

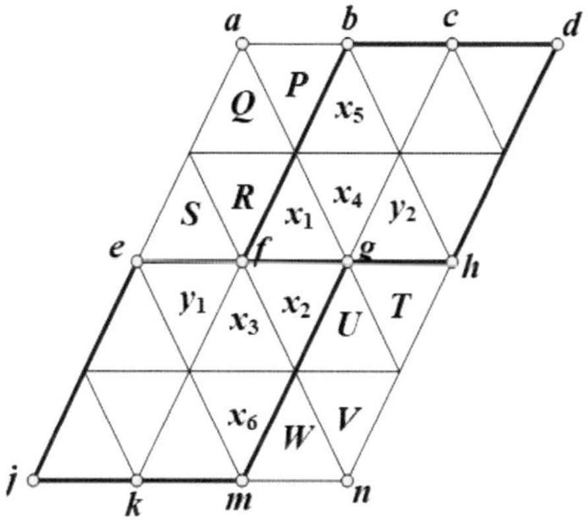

Figure 10.6

We have just shown that parallelogram (adjn) in Figure 10.6 has value 0. Parallelogram (acjm) has a Δ_2 dissection so it also has value 0. By subtracting we obtain ν(cdmn) = 0, and similarly ν(abjk) = 0 and ν(bckm) = 0. Thus a P(3,4) parallelogram has a dissection into three P(1,4) parallelograms all with value 0. These constituent parallelograms are relatively nilpotent; even in a P(2,4) parallelogram the constituent P(1,4) parallelograms need not be nilpotent.

As an immediate Corollary we obtain

Corollary 10.8 *The H(2,2,1) hexagon is nilpotent.*

Proof. The hexagon (abhnme) in Figure 10.6 is H(2,2,1) and the proof in Theorem 10.7 that P+Q+R+S+T+U+V+W = 0 also shows that this hexagon is nilpotent. □

Corollary 10.9 *The parallelogram P(4t,q) is nilpotent for every positive integer t and for every integer q ≥2.*

Proof. To the P(3,4) in Figure 10.6 any number of P(1,4) parallelograms can be appended, each one with value 0, to obtain a nilpotent P(4, q) for any integer q ≥3. Any number of P(4,q) parallelograms can then be appended *below* to obtain the desired P(4t,q) nilpotent parallelogram. □

Theorem 10.10 *If a, b are both odd then P(a,b) is not nilpotent.*

Proof. Our usual method of constructing a locally nilpotent assignment using the linear pattern over {0, 1, -1} works equally well here. In Figure 10.7 it can be seen that the linear pattern gives a value of 1 to the P(3,3) parallelogram (abcd). Thus P(3,3) is not nilpotent. The appended parallelograms (bedf) and (cfgh) each have value 0. Such 0- value parallelograms can be repeatedly appended to obtain a P(a,b) parallelogram with value 1 for any pair of odd values a,b each greater than or equal to 3. Clearly if a = 1 or b = 1, then P(a,b) is not nilpotent. □

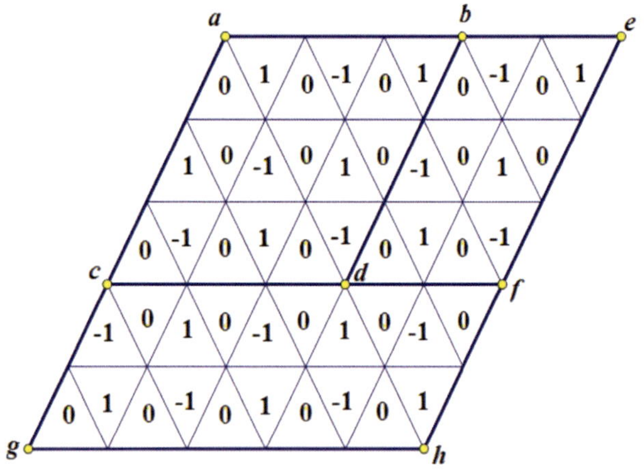

Figure 10.7

Theorem 10.11 *The triangulated parallelogram P(a,b) is nilpotent iff (i) a, b are both even, or (ii) one of the parameters, say a, is a multiple of 4, and b is any integer larger than or equal to 2.*

Proof. It remains only to show that P(4t + 2, q) is nonnilpotent for odd q and any positive integer t. To this end consider Figure 10.8. The parallelogram in this Figure is locally nilpotent and can be extended uniquely to a locally nilpotent triangular tiling of the entire plane. The stars indicated in bold and the Δ_2 triangles are composed entirely of 0-valued triangles although we have not entered the 0 values. This is a second type of pattern over {0,1,-1} which we call a *star and triangle* pattern.

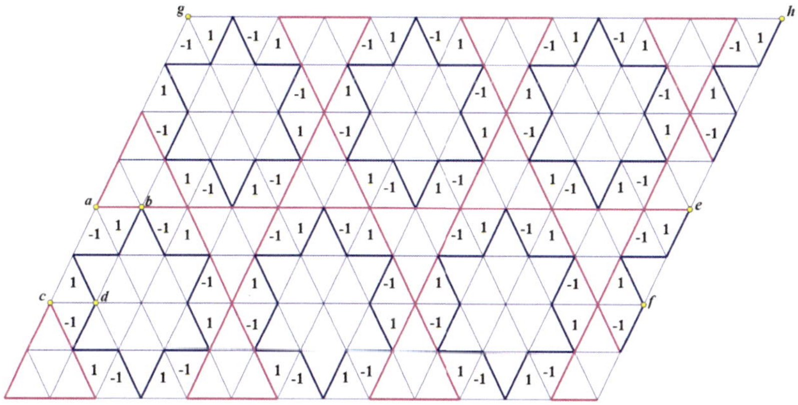

Figure 10.8

By Corollary 10.9 the parallelogram (ghae) is nilpotent. Also parallelogram (bedf) is nilpotent since it has a Δ_2 dissection. The parallelogram (abcd) has value 1, and thus the parallelogram (ghcf) has value 1, showing that P(6,13) is nonnilpotent. If Figure 10.8 is studied closely, it is seen that P(6, q) has value 1 for any odd integer q≤13. Also the continued unique extension obtained by appending P(6,2) parallelograms shows that P(6,q) is nonnilpotent for any odd integer q. The P(6,q) parallelogram can be topped by any number of P(4,q) parallelograms all with value 0. The end result is a P(4t+2, q) parallelogram whose value remains at 1. □

10.5 Nilpotent Triangulated Pentagons

A nilpotent convex pentagon tiled with equilateral triangles has two pairs of parallel sides and one interior acute angle as in the examples in Figure 10.9.

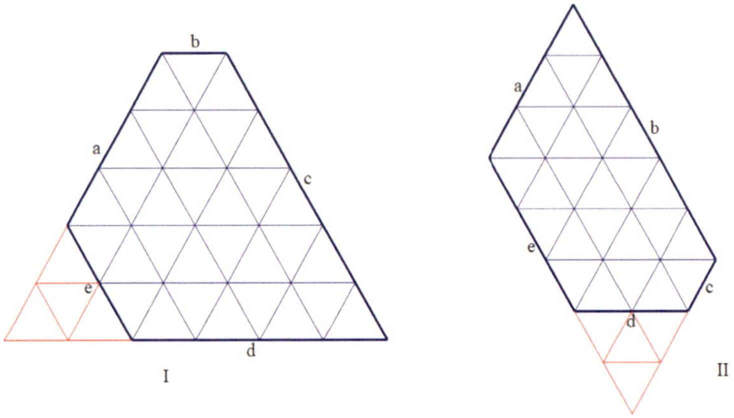

Figure 10.9

In Figure 10.9, a, b, c,d and e represent the lengths of the sides. In 10.9(I) a,b,c represent odd lengths. Since a+b = d and a+e = c, both e, d must be even. Thus the indicated Δ_e can be appended to form a trapezoid. The lateral sides of this trapezoid do not have length a multiple of 4, so by Theorem 10.6 it is nonnilpotent. Since the appended Δ_e is nilpotent, it must be that the original pentagon is nonnilpotent.

In Figure 10.9(II), a, b,c, e represent odd lengths. Since c+d = a, it must be that d is even. We have shown that every triangulated pentagon has at least one side with even length. In 10.9(II) the appended Δ_e creates a P(3,5) parallelogram which is nonnilpotent by Theorem 10.10. It follows again that the original pentagon is nonnilpotent. We have proved the following result.

Theorem 10.12 *Let P be a pentagon tiled with equilateral triangles. Then P has at least one side with even length e. The nilpotent triangle* Δ_e *can be appended to P along this side to create either a parallelogram* *or an isosceles trapezoid. If these figures are nonnilpotent, then P is nonnilpotent, and if these figures are nilpotent, then so is P.* □

10.6 Nilpotent Triangulated Regular Hexagons

In case H is a hexagon tiled with equilateral triangles, then unlike Theorem 10.12, all six sides of H may have odd length. If however H does contain a side with even length then the same kind of operation described in Theorem 10.12 can be performed and the nilpotence or nonnilpotence of H will be aligned with the nilpotence or nonnilpotence of a polygon with fewer than 6 sides. We will for the time being focus our attention on hexagons in which all six sides have odd length.

We start by considering regular hexagons with an odd number of edges per side. The notation H_m will be used to denote a regular hexagon with m edges per side. In Figure 10.10 dissections of H_3 and H_5 are illustrated using *stars*, H_1 hexagons and Δ_2 triangles. The six -pointed star will be denoted by J_1. The components of these dissections, including J_1, are nilpotent, and thus H_3 and H_5 are themselves nilpotent.

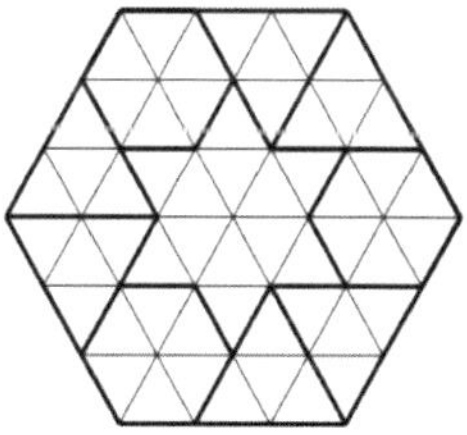

I

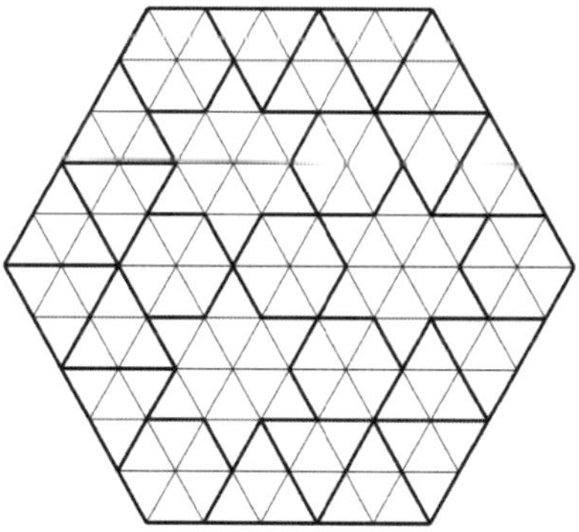

II

Figure 10.10

The constructions in Figure 10.10 illustrate a regular formation consisting of stars and hexagons. This formation can be used to obtain a $\{J_1, H_1, \Delta_2\}$ dissection of H_m for any odd m. The tiling of

H_7 will consist of 6 stars arranged point to point in a triangular formation. Since also H_m has a Δ_2 dissection if m is even, we have:

Theorem 10.13 *The regular hexagon H_m is nilpotent for every positive integer m.* □

A nilpotent polygon P which has no dissection into smaller intrinsically nilpotent polygons, will be called *prime,* and if P does have such a dissection, it will be termed *composite.* Δ_2, H_1, J_1, H(2,2,1) are *simple* prime polygons.

The star and hexagon phalanx is not the only way to obtain a dissection of H_m into prime nilpotent components. In Figure 10.11 two such dissections of H_5 are illustrated. They feature the use of another simple prime polygon we call a *link* since they are generally used to link two H_1 hexagons thus forming a nilpotent H(1,1,5). The links in Figure 10.11 are labeled with an L. In Figure 10.11(I) note the appearance of two prime H(1,2,2) hexagons, which along with two Δ_2 triangles, form dissections of T(1,4,5) trapezoids. In Figure 10.11(II) note the inclusion of an H_3 hexagon inside the H_5. This inclusion will be denoted by $H_3 \rightarrow H_5$ meaning that the dissection of H_5 into basic nilpotent figures contains a similarly dissected H_3.

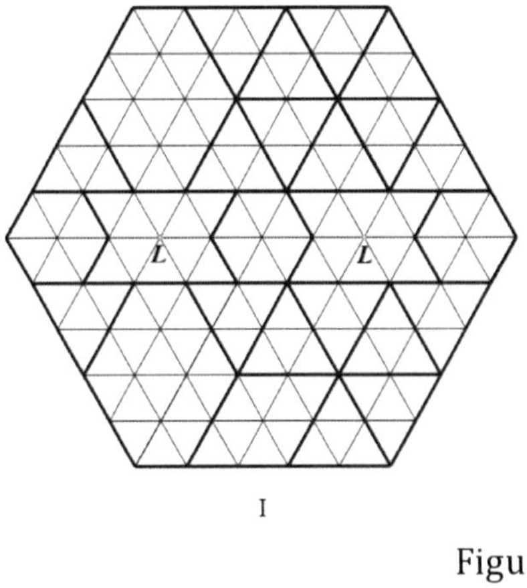

I

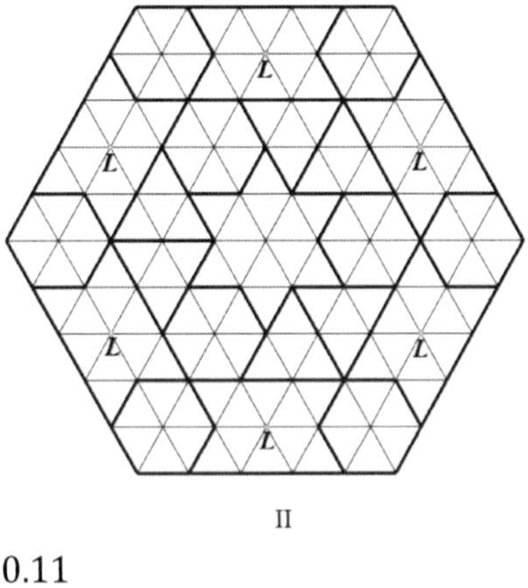

II

Figure 10.11

In Figure 10.12 the extension $H_3 \rightarrow H_7$ is achieved. This is an example of a *simple extension.* A simple extension is one which involves only the simple prime polygons Δ_2, H_1, J_1, L, H(2,2,1). Similar extensions involving larger regular hexagons are also possible.

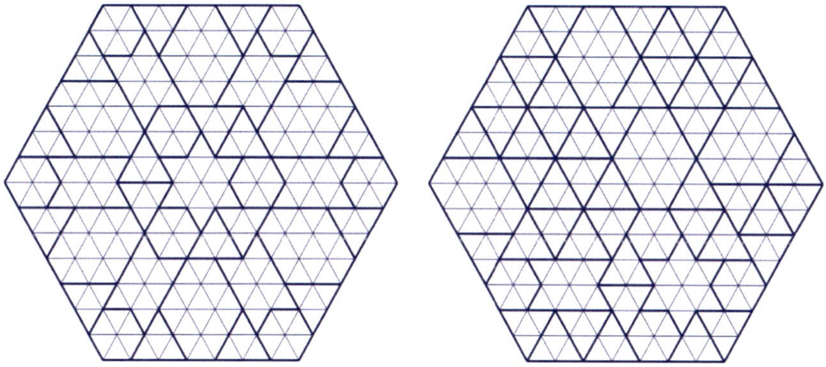

Figure 10.12

In the $H_3 \rightarrow H_7$ extension of Figure 10.12 at left there can be found three P(6,6) parallelograms which could have been just given Δ_2 dissections. At right in Figure 10.12 another $H_3 \rightarrow H_7$ extension is illustrated this time with H_3 occupying a corner position. In Figure 10.13 the extension $H_3 \rightarrow H_5 \rightarrow H_7$ is obtained with H_3 occupying a corner of H_5 and H_5 occupying a corner of H_7.

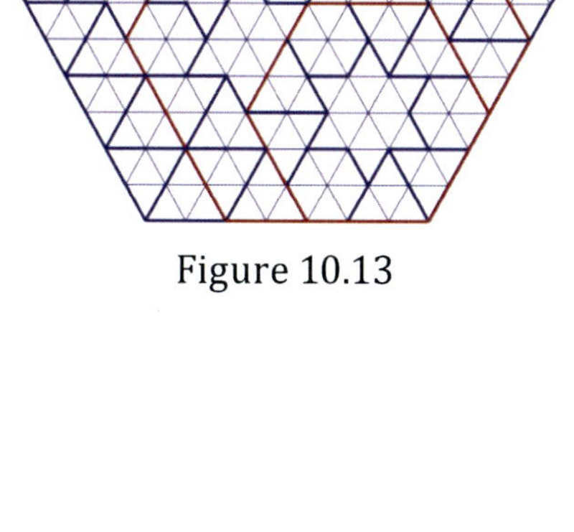

Figure 10.13

10.7 Semiregular Nilpotent Hexagons

A semiregular hexagon of the form H(a,b,b) with a≠b is *near-regular.* We investigate near-regular hexagons starting with a = 1. In Corollary 10.8 it was shown that H(1,2,2) is nilpotent. It is easily determined that H(1,3,3) is not nilpotent. Figure 10.14 illustrates the composite nature of H(1,4,4), H(1,5,5) and H(1,6,6) and thus these hexagons are nilpotent.

A composite construction of triangulated hexagons of the form H(1,q,q) where q is even can be obtained by stacking q/2 H(1,2,2) hexagons and completing the dissection with Δ_2 triangles, as in 10.14(I,III). If q is odd and has the form q = 4t+1, then H(1,q,q) has a dissection consisting of the parallelogram P(4t,4t) encased on one side by two copies of H(1,1,4t+1) which have an H_1 in common, as in 10.14(II). Note that nilpotent H(1,1, 4t+1) can be obtained for any positive integer t by connecting H_1 hexagons by links.

Finally, we leave it as an exercise to show that if q = 4t+3, then H(1,q,q) is *not* nilpotent. Hint: show that H(1,3,3) is not nilpotent and that for t>1, H(1,q,q) can be tiled by means of a nilpotent extension of H(1,3,3). Here it must be shown that there exists an integer assignment to H(1,3,3) so that ν(H(1,3,3))≠0 but which can be extended to an integer assignment to H(1, 4t+3, 4t+3) which is locally nilpotent outside of H(1,3,3). Questions involving free generating sets and locally nilpotent assignments will be considered in Chapter 12.

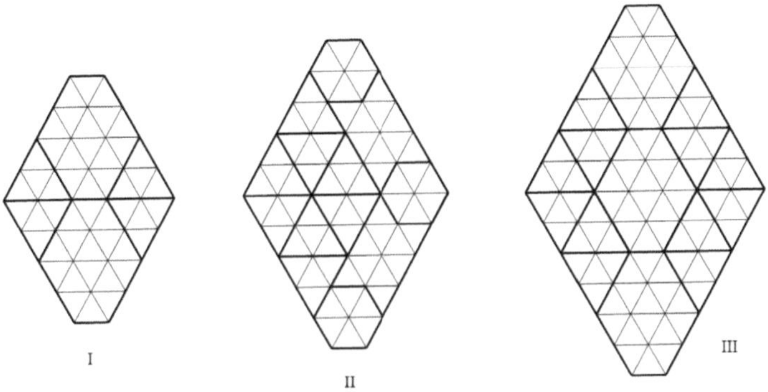

Figure 10.14

We thus have the following result.

Theorem 10.14 *The triangulated hexagon H(1,q,q) is nilpotent except when q has the form q = 4t+3.* □

In case one of the parameters is even we have the following result.

Theorem 10.15 *If p is even then the semiregular hexagon H(p,q,r) is nilpotent iff (i)q,r are both even or (ii) one of the parameters say q is even and r is odd and p+q is a multiple of 4.*

Proof. If p,q,r are all even then H(p,q,r) has a Δ_2 dissection and thus is nilpotent. If p,q are even; r is odd and p+q is a multiple of 4, then the extension obtained by appending a Δ_p to opposite sides of length p is the parallelogram P(p+q, r). By Theorem 10.11 this parallelogram is nilpotent. It then follows that H(p,q,r) is also nilpotent. If in this case p+q is not a multiple of 4, then the parallelogram P(p+q, r) obtained by extension is not nilpotent by Theorem 10.11. As a consequence H(p,q,r) is also not nilpotent. □

As a consequence of Theorem 10.15, we can focus on semiregular hexagons H(p,q,r) in which the parameters p,q,r are all odd.

Theorem 10.16 *Let H(p,q,r) be a semiregular hexagon with p,q,r all odd. Then H(p,q,r) is nilpotent iff (i) each of p,q,r is congruent to 1 (mod 4), or (ii) each of p,q,r is congruent to 3 (mod 4).*

Proof. Figure 10.15 illustrates the extension H(p,q,r)→H(p+4s,q,r) by means of parallelograms P(4s,r) and P(4s,q). By Corollary 10.9 each of these parallelograms is nilpotent and thus H(p,q,r)→H(p+4s,q,r) is a simple extension. As a consequence H(p+4s,q,r) is nilpotent iff H(p,q,r) is; moreover H(p,q,r) is nilpotent iff H(p′,q′,r′) is, where p′ = p (mod 4); q′ = q (mod 4) and r′ = r (mod 4). The nilpotence of H(p,q,r) thus depends on the nilpotence of H_1 = H(1,1,1), H(1,1,3), H_3 = H(3,3,3) or H(1,3,3). The hexagon H_1 is nilpotent by definition, and we have seen in Figure 10.10(I) that H_3 is composite, hence nilpotent. On the other hand H(1,1,3) and H(1,3,3) are nonnilpotent and so any nilpotent extension of them is also nonnilpotent. □

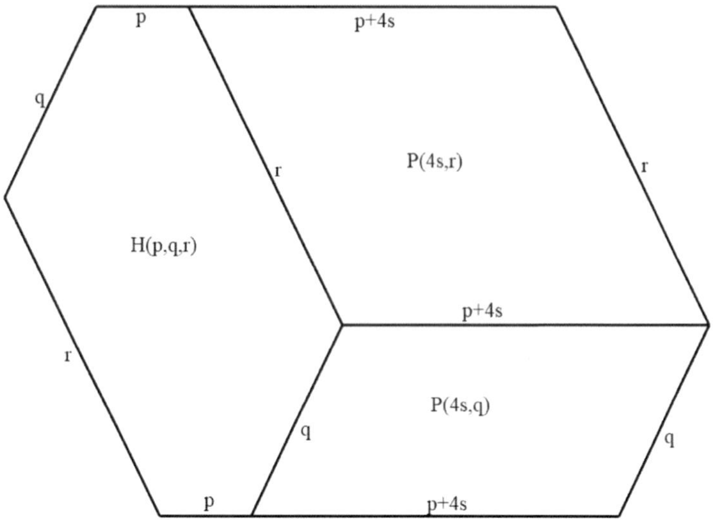

Figure 10.15

10.8 Prime Semiregular Hexagons

From Theorem 10.16 it is seen that every nilpotent semiregular hexagon H(p,q,r) with p,q,r all odd is a simple extension of H_1, and thus H_1 is the only prime hexagon with all sides of odd length. If p,q,r are all even, then H(p,q,r) has a Δ_2 dissection and thus is composite. In case one parameter, say p, is even and q,r are both odd, then H(p,q,r) can be extended to the parallelogram P(q+p, r+p) by appending two copies of the nilpotent triangle Δ_p. But this parallelogram is nonnilpotent by Theorem 10.10, and it follows that H(p,q,r) is also nonnilpotent. Even if p,q are even, r is odd and p+q is *not* a multiple of 4, then H(p,q,r) is nonnilpotent by Theorem 10.11

In case p, q are even, r is odd and p+q is a multiple of 4, we have the following result.

Theorem 10.17 *H_1 and H(2,2,1) are the only prime semiregular hexagons.*

Proof. It remains only to show that if p,q are even, r is odd and p+q is a multiple of 4, then H(p,q,r) can be constructed as a simple extension of H(2,2,1). In Figure 10.16 it is illustrated how a hexagon of the form H(4s+2, 6+4k, 2t+1) can be obtained via such an extension.

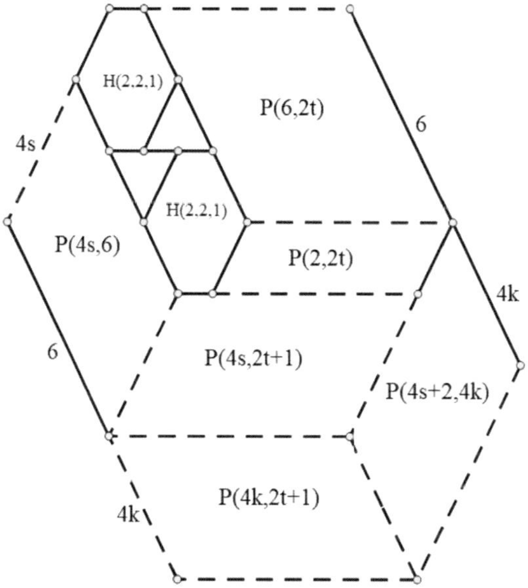

Figure 10.16

A second possibility occurs when both even sides are multiples of 4, as in Figure 10.17. In 10.17(I) it is shown that the hexagon H(2t+1, 4s+4, 4k) is a nilpotent extension of H(2,2,1) for s,k larger than or equal to 1 and t ≥2. (In case t = 2, the trapezoid T(2t-4,4,2t) reduces to the triangle Δ_4 and in case s = 0 the hexagon H(2t+1,4,4k) is obtained. In Figure 10.17(II) the case t = 1 is illustrated. In all cases H(p,q,r) is obtained as a simple extension of H(2,2,1). □

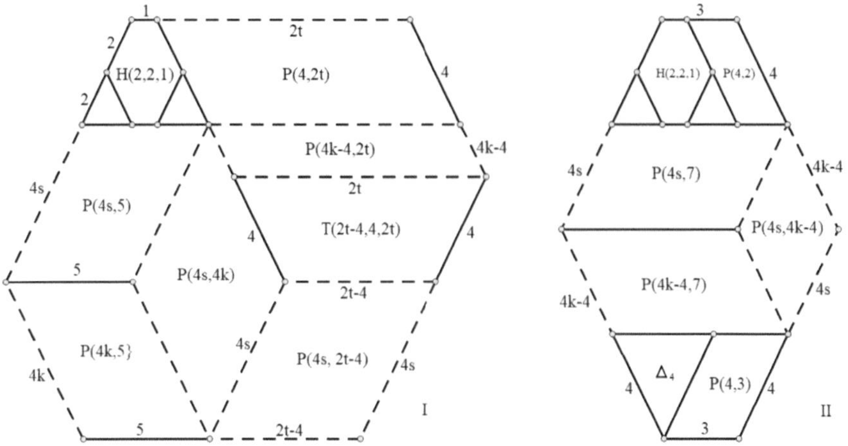

Figure 10.17

10.9 Nilpotent Nonsemiregular Hexagons with all Sides Odd

A general convex triangulated hexagon will be denoted by H(p,q,r,s,t,u) where the six parameters represent the lengths of sides in order following a clockwise traversal of the boundary. Opposite sides will be parallel but may differ in length. The equations p+q = s+t and r+q = t+u will always hold. The only nonsemiregular triangulated quadrilaterals are the isosceles trapezoids covered in Section 10.3, and triangulated pentagons covered in Section 10.5 are necessarily nonsemiregular.

In this Section we confine our attention to nonsemiregular hexagons each of whose sides has odd length. Two important reductions are indicated by the following.

Lemma 10.18 *Let H(p,q,r,s,t,u) be a nilpotent triangulated hexagon with all parameters odd. If any two opposite sides, say p, s, are each augmented by 4 units then the resulting hexagon H(p+4,q,r,s+4,t,u) is also nilpotent. Any three consecutive sides, say p,q,r, with q≥5, can be replaced by p+4, q-4,r+4 creating the nilpotent hexagon H(p+4, q-4, r+4, s,t,u).*

Proof. In Figure 10.18(I) the first extension is indicated and in Figure 10.18(II) the second extension is illustrated. □

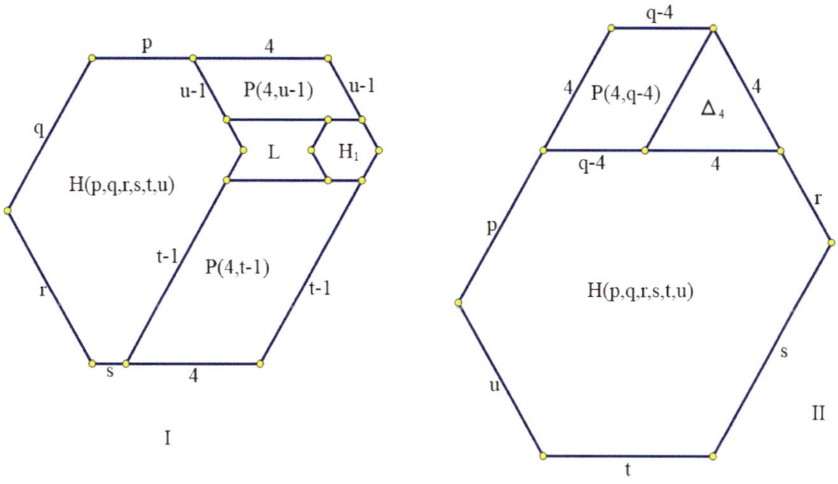

Figure 10.18

A hexagon of the form H(a,b,a,b,a,b) with a≠b will be called *alternating*. We will study these hexagons first.

Theorem 10.19 *The alternating hexagon H(a,b,a,b,a,b) with a, b odd, is nilpotent iff a+b = 4t+2 for some positive integer t.*

Proof. In Figure 10.18 a nested construction indicates a procedure for obtaining every alternating hexagon with b = a+4 as an extension of H_1. The illustrated extension sequence is H_1 = H(1,1,1,1,1,1) → H(1,1,5,1,1,5) → H(1,5,1,5,1,5) → H(3,7,3,7,3,7) → H(5,9,5,9,.5,9) → H(7,11,7,11,7,11). Basically, there is a simple extension H(a,b,a,b,a,b)→H(a+2,b+2,a+2,b+2,a+2,b+2).

If a = 4q+1 and b=4r+1 then H(a,b,a,b,a,b) can be reduced to a hexagon of the form H(1,s,1,s,1,s), s = 4t+1 by subtracting 2 repeatedly from a and b. If s = 1, then H(1,s,1,s,1,s) = H_1 and it follows that H(a,b,a,b,a,b) is nilpotent. Otherwise, if s≥5 then the reductions H(1,s,1,s,1,s)→H(1,s-4,5,s-4,1,s) → H(1,s-4,1,s-4,1,s-4) are possible due to Lemma 10.18. This reduction process can be repeated until H(1,1,1,1,1,1) = H_1 is obtained, again with the consequence that H(a,b,a,b,a,b) is nilpotent.

Now suppose a =4q+1 and b = 4r+3. Then H(a,b,a,b,a,b) reduces to a hexagon of the form H(1,s,1,s,1,s) with s = 4t+3. If s = 3 then H(1,s,1,s,1,s) = H(1,3,1,3,1,3) which is nonnilpotent, and H(a,b,a,b,a,b) is likewise nonnilpotent. Otherwise, if s≥7 the nilpotent reductions H(1,s,1,s,1,s) →H(1,s-4,5,s-4,1,s)→H(1,s-4,1,s-4,1,s-4) can be performed. By repeating this reduction rocess, ultimately H(1,3,1,3,1,3) is obtained, again with the consequence that H(a,b,a,b,a,b) is nonnilpotent.

Finally, if a = 4q+3 and b = 4r+3, then H(a,b,a,b,a,b) reduces to a hexagon of the form H(3,s,3,s,3,s) with s = 4t+3. If s = 3 then H(3,s,3,s,3,s) = H_3 which is nilpotent. Consequently H(a,b,a,b,a,b) is also nilpotent. Otherwise, if s≥7, then the reductions H(3,s,3,s,3,s) → H(3,s-4,7,s-4,3,s) → H(3,s-4,3,s-4,3,s-4) can be performed. By repeating this reduction process, ultimately the hexagon H_3 is again obtained, with the consequence that H(a,b,a,b,a,b) is nilpotent. □

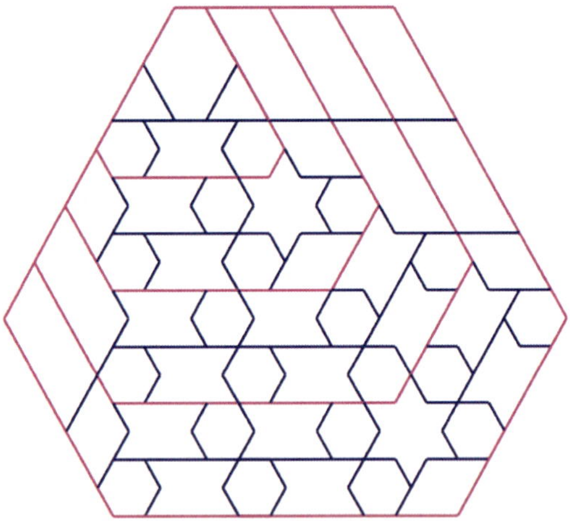

Figure 10.18

To determine the genesis of hexagons H(p,q,r,s,t,u) with all parameters odd, it is necessary to determine the *irreducible* hexagons in this class. By *irreducible,* we mean those hexagons for which the reductions described in Lemma 10.18 are not applicable. These hexagons are described in the following result.

Theorem 10.20 *The irreducible hexagons with all odd parameters consist precisely of H_1, H_3, H(1,3,1,3,1,3), H(1,1,3,1,1,3), H(3,3,1,3,3,1) and H(5,1,3,3,3,1).*

Proof. Suppose the hexagon H(p,q,r,s,t,u) is irreducible with all odd parameters, and suppose one of the parameters, say p, is larger than or equal to 5. Then it must be that the opposite side, (s) has length 1 or 3, since otherwise the first reduction of Lemma 10.18 could be applied. If q≥5, then similarly, t ∈ {1,3}. But the requirement p+q = s+t cannot then be satisfied since p+q ≥10 and s+t≤6. It must then be that q∈ {1,3}. If r ≥5, then H(p,q,r,s,t,u) → H(p-4, q+4, r-4, s,t,u) is a possible reduction, so it must be that r ∈{1,3}. If u≥5 then u+p≥10, but u+p = r+s ≤6, so we must have u ∈{1,3}. Fonally, if t≥5 then H(p,q,r,s,t,u) → H(p-4,q,r,s,t-4,u+4) is a possible reduction.

The conclusion is that if p≥5, then the remaining four parameters must all be 1 or 3. Since p+q = s+t ≤6 it must be that p = 5 and q = 1, and since p+u = r+s ≤6, it must be that u = 1. Then p+q = 6 = s+t yields s=t=3, and q+r = t+u forces r = 3. Thus H(5,1,3,3,3,1) is the only irreducible hexagon which contains a parameter with value larger than or equal to 5.

If H contains two consecutive parameters, say p, q with value 1, then p+q = s+t forces s=t=1. Then either r=u=3 resulting in H(1,1,3,1,1,3) or r=u=1 resulting in H_1. If H contains two consecutive parameters, say p, q, with value 3, then p+q = s+t forces s = t =3. Then either r=u=1 resulting in H(3,3,1,3,3,1), or r=u=3 resulting in H_3. If no two consecutive parameters are equal, then of course the hexagon H(1,3,1,3,1,3) results. □

In Figure 10.19 is introduced another pattern, the *wedge and rhomb pattern over {0,1,-1}*. It is generated by the star and adjacent triangle outlined in red in 10.19(I). As with the linear and star and tiangle patterns, the wedge and rhomb pattern provides a simple way to obtain a locally nilpotent assignment which extends to a triangulation of the entire plane or any polygonal subregion of the plane. The irreducible hexagons H(1,3,1,3,1,3), H(1,1,3,1,1,3), H(3,3,1,3,3,1) and H(5,1,3,3,3,1) determined in Theorem 10.20, have been superimposed in the wedge and rhomb patterns in 10.19(I, II). Each superimposition has been chosen so that the value of these hexagons is nonzero. Any nilpotent extension of these will then have nonzero value in the wedge and rhomb pattern, and thus be nonnilpotent.

The reductions given in Lemma 10.18 always either increase or decrease parameters by 4 units. Thus if p is a parameter then p+4, p-4 and p are all either congruent to 1 (mod 4) or congruent to 3 (mod 4). If the hexagon H = H(p,q,r,s,t,u) contains one parameter which is congruent to 1 (mod 4), and a second parameter which is congruent to 3 (mod 4), then H must reduce to one of the four nonnilpotent hexagons listed in Theorem 10.20, and consequently H itself must be nonnilpotent. On the other hand, if every parameter of H is congruent to 1, then H reduces to H_1, and if every parameter of H reduces to 3 then H reduces to H_3. In both cases H will be composite and thus nilpotent. We have proved the following.

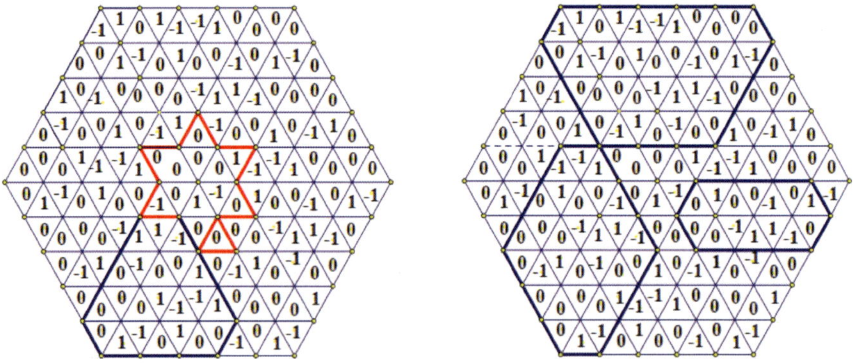

I II

Figure 10.19

Theorem 10.21 *If H = H(p,q,r,s,t,u) is a triangulated convex hexagon with all parameters odd, then H is nilpotent iff all parameters are* *congruent to 1 (mod 4), or all parameters are congruent to 3 (mod 4). In the first case H is a simple extension of H_1, and in the second case H is a simple extension of H_3. Since H_3 is itself composite, it follows that H_1 is the only prime hexagon with all sides odd.* □

10.10 Nilpotent Hexagons with at Least one Even side

If the hexagon H = H(p,q,r,s,t,u) has all even sides then it has a Δ_2 dissection and thus is nilpotent. So we will consider in this Section hexagons with at least one even side and one odd side. The reductions given in Lemma 10.18 apply here also, and the same argument given in Theorem 10.20 can be applied to obtain: if p≥7 then the remaining parameters q,r,s,t,u must all lie in the range {1,2,3,4}. If p = 7 then p+q = s+t forces q = 1 and s = t = 4, so we have H(7,1,r,4,4,u). Now since u+7 = r+4 we must also have u = 1 and r = 4 and we obtain the hexagon H(7,1,4,4,4,1) as the only irreducible hexagon with one parameter equal to 7. Appending two copies of Δ_4 to sides r, t yields an isosceles trapezoid with lateral sides of length 5. By Theorem 10.6, H is nonnilpotent.

Now suppose H has a side with length 6, say p = 6. In this case there are only three irreducible hexagons: H(6,2,3,4,4,1), H(6,1,4,3,4,1) and H(6,1,3,4,3,1). Two Δ_4 triangles can be attached to sides r, t of H(6,1,4,3,4,1) to obtain an isosceles trapezoid with lateral sides of length 5. By Theorem 10.6, such a trapezoid is nonnilpotent, and it follows that H is also nonnilpotent. A Δ_4 and a Δ_6 triangle can be attached one to side p and the other to side s of H(6,1,3,4,3,1) to obtain a parallelogram with each side of length 7. By theorem 10.11 such a parallelogram is nonnilpotent, and it follows that H also is nonnilpotent.

The hexagon H(6,2,3,4,4,1) is both nilpotent and composite as illustrated in Figure 10.20. In Figure 10.20(I) note the extensions H(1,2,2)→H(1,2,6)→H(6,2,3,4,4,1), and H(1,2,2)→H(2,2,3)→H(6,2,3,4,4,1) and in Figure 10.20(II) the extension H(1,2,2)→ H(1,4,4) →H(6,2,3,4,4,1). Besides H(6,2,3,4,4,1) no nonsemiregular irreducible hexagons can be found. As a consequence we have the following improvement of Theorem 10.17.

Theorem 10.22 H_1 and H(2,2,1) are the only prime triangulated convex hexagons. □

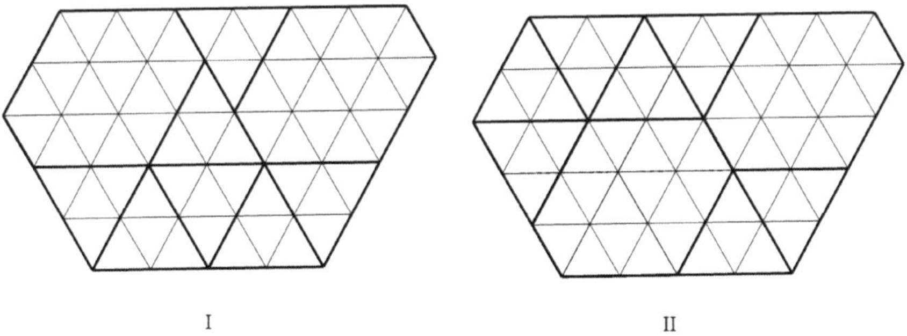

Figure 10.20

10.11 Exercises

(1) If q = 4t+3, show that the near regular hexagon H(1,q,q) is nonnilpotent.

(2) Show that the (i) linear; (ii) star and triangle, and (iii) wedge and rhomb are the only patterns over {0,1,-1} that make a locally nilpotent assignment for the entire plane.

(3) Let P be a triangulated polygon, and let R_1, R_2 be two congruent subpolygons of P. If for any locally nilpotent assignment of P, the regions R_1, R_2 have exactly the same values assigned to corresponding unit triangles, then R_1 *is equivalent to* R_2. For the triangulated polygon in Figure 10.21, show that the three indicated corner hexagons are equivalent. Also show that the region R obtained by removing the three corner hexagons is prime. R is an example of a *nonsimple* or *complex prime.*

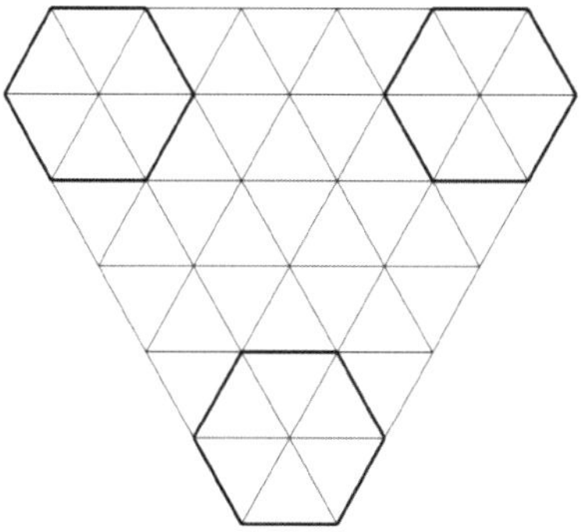

Figure 10.21

4. Let Ω denote a triangulation of the entire Euclidean plane. Superimpose on Ω a tiling by P(4,4) parallelograms as in Figure 10.22. Show that all the P(4,4) parallelograms in this tiling are equivalent. How many unit triangles lie in a free generating set (FGS) for Ω?

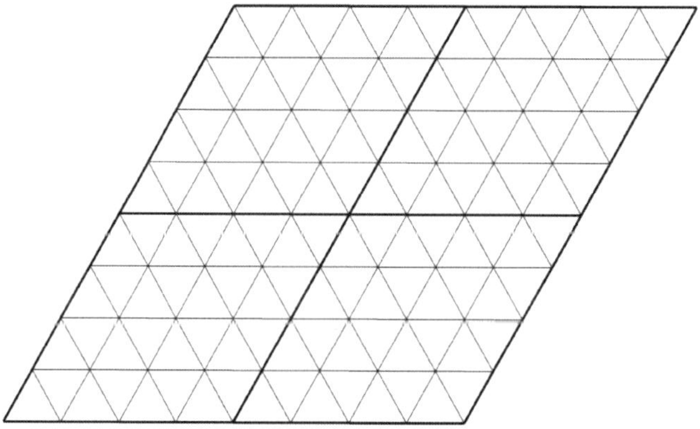

Figure 10.22

5. Determine free generating sets for triangulated (i) triangles; (ii) trapezoids; (iii) parallelograms, and (iv) hexagons. Which general triangulated polygons have an FGS consisting of only six unit triangles?

CHAPTER 11

Nilpotent Triangulated Stars

11.1 Regular Stars

Let J_m denote the triangulated star shaped polygon obtained by attaching triangles Δ_m to each side of the regular hexagon H_m. We call J_m the *regular star with side m.* In case m is even then J_m has a Δ_2 dissection and consequently is nilpotent. We shall therefore confine our attention in this Section to stars J_m with m odd. The simple star J_1 was introduced in Chapter 10 where it functioned as a prime component in dissections of regular hexagons. The star J_3 is illustrated in Figure 11.1.

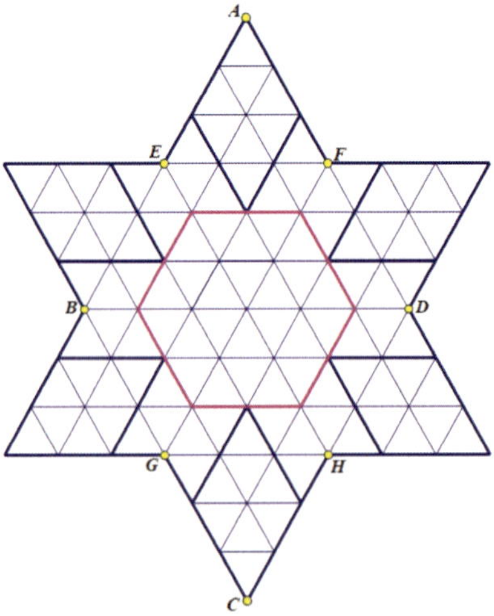

Figure 11.1

In Figure 11.1 J_3 is dissected using six P(2,2) parallelograms and a polygon formed by attaching six *wedges* to the corners of an H_2 hexagon. We denote this *6-wedge* by $[H_2 : W^6]$ and leave it as an exercise to show that it is a (complex) prime.

Lemma 11.1 *The 6-wedge $[H_2 : W^6]$ is prime.* □

It is easily seen that J_m is nilpotent for every positive integer m.

Theorem 11.2 *The star J_m is nilpotent for every positive integer m.*

Proof. When two copies of Δ_m are attached to opposite sides H_m, the parallelogram P(2m,2m) is obtained. This parallelogram is nilpotent by Theorem 10.11. Since H_m is nilpotent by Theorem 10.13, it follows that the sum of the values of the attached Δ_m triangles is 0. As a consequence, the sum of the three pairs of Δ_m triangles attached to opposite sides of H_m is 0. But this is how J_m is formed, and thus J_m is nilpotent. □

With regard to the composite or prime character of J_m we have the following result.

Theorem 11.3 *The star J_1 is prime. If m is even then J_m is composite. If m≠1,3 then J_m can be constructed as a nilpotent extension of H_{m-2} and thus is composite. J_3 can be dissected as in Figure 11.1 using the prime 6-wedge and six P(2,2) parallelograms, and thus J_3 is also composite.*

Proof. The star J_1 is an important prime building block. The composite construction of J_3 is indicated in Figure 11.1. If m is even then J_m has a dissection consisting entirely of Δ_2 triangles. So suppose m≥5 is odd.

Case i: m = 4t+1. The general construction for this Case is illustrated in Figure 11.2.

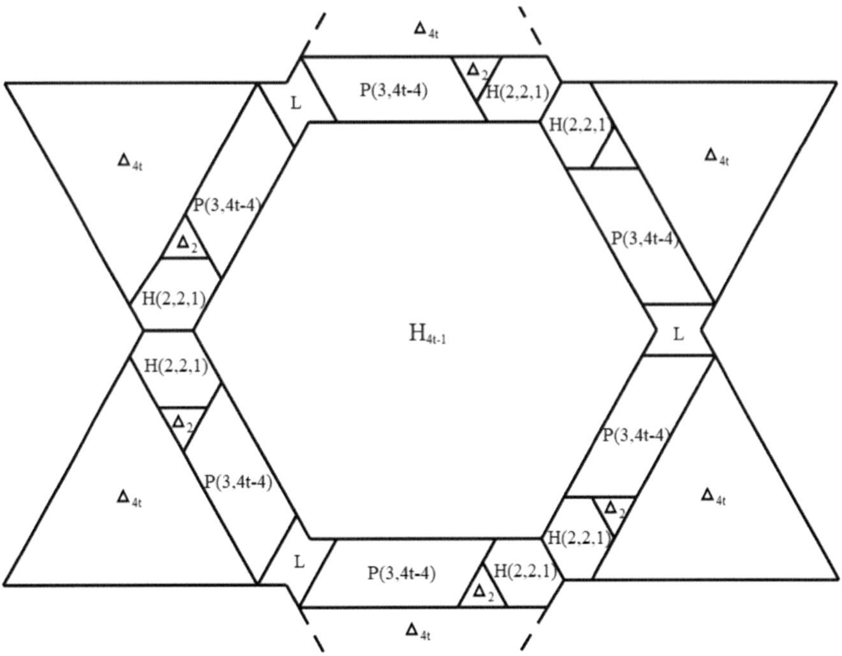

Figure 11.2

Case ii: m = 4t+3. For the purposes of illustration, let L^k denote the polygon consisting of k links in a row separated by k-1 H_1 hexagons. The general construction for this Case is given in Figure 11.3. □

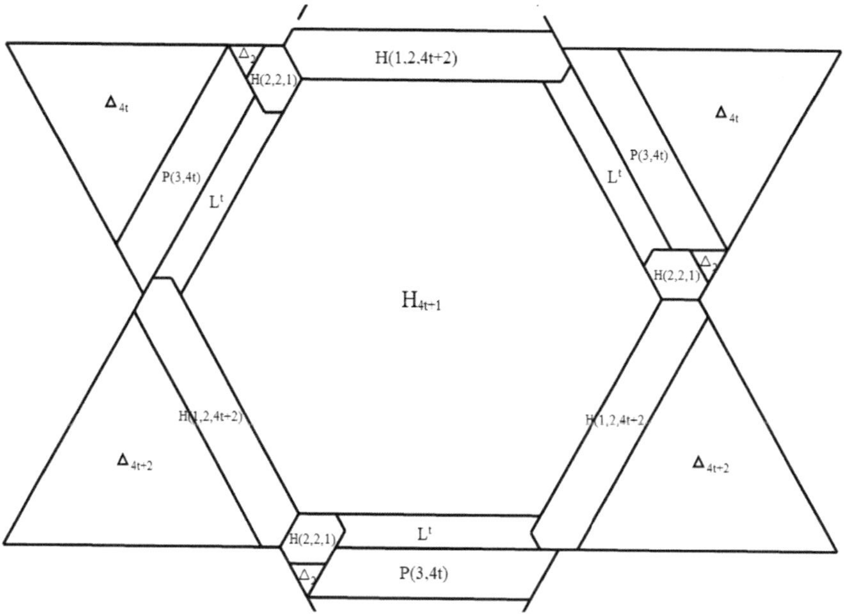

Figure 11.3

11.2 Irregular Stars

Starting with a triangulated hexagon H_m with m odd, besides attaching copies of Δ_m to each side, we attach copies of Δ_p, (p odd and less than m), to the center of each side, we obtain an *irregular star.* We denote this star by $[H_m : \Delta_p]$. In Figure 11.4 the irregular star $[II_3 : \Delta_1]$ is illustrated.

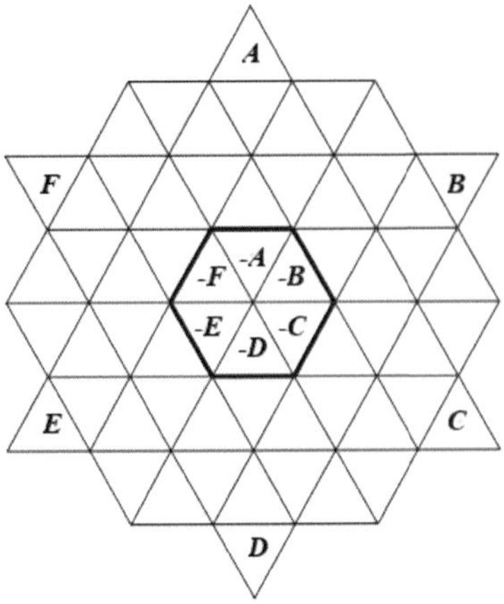

Figure 11.4

In Figure 11.4 note that the attached Δ_1 triangles have opposite values which fill the central H_1 hexagon. These values then sum to 0, with the consequence that $[H_3 : \Delta_1]$ is nilpotent. Also $[H_3 : \Delta_1]$ has no dissection into smaller prime figures, and thus is itself a (complex) prime. In general we have the following result.

In $[H_m : \Delta_p]$ suppose $m = p + 2q$ with p odd and q even as in Figure 11.5.

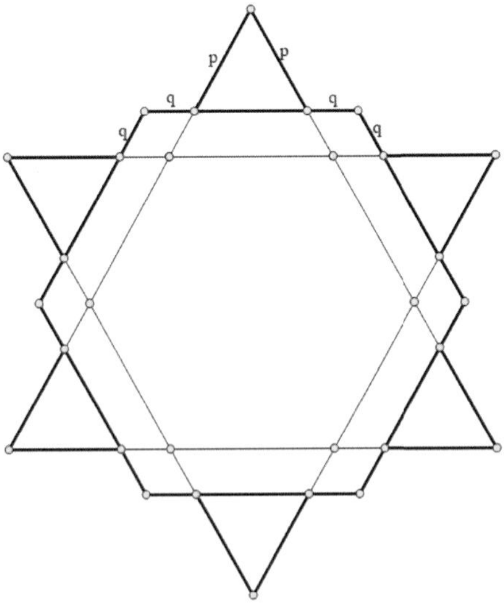

Figure 11.5

Note that $[H_m : \Delta_p]$ is just an extension of J_{p+q} by means of six P(q,q) parallelograms. Thus the composite nature of $[H_m : \Delta_p]$ follows from the composite nature of J_{p+q}. We can then suppose that q is odd. In Figure 11.6 the irregular star $[H_{4t+7} : \Delta_{4t+1}]$ is shown to be a composite extension of $[H_7 : \Delta_1]$, which in turn, is an extension of the prime $[H_3 : \Delta_1]$. We consider the following Cases.

Case i : p = 4t+1 and m = p+2q with q = 4u+1. In order to illustrate a composite construction in this Case we define a *star chain*. In Figure 11.7 it is seen that a star chain consists of figures consisting of a J_1 star and two adjacent H_1 hexagons, connected together in linear sequence. If such a sequence contains k stars, we denote it by S^k. In Figure 11.8, the general construction for this Case is illustrated, where it must be supposed that t, u are each larger than or equal to 1.

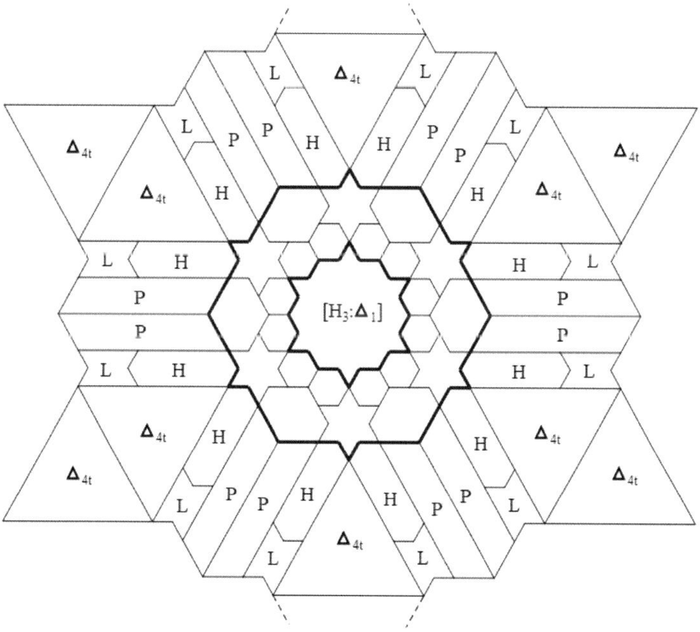

Figure 11.6

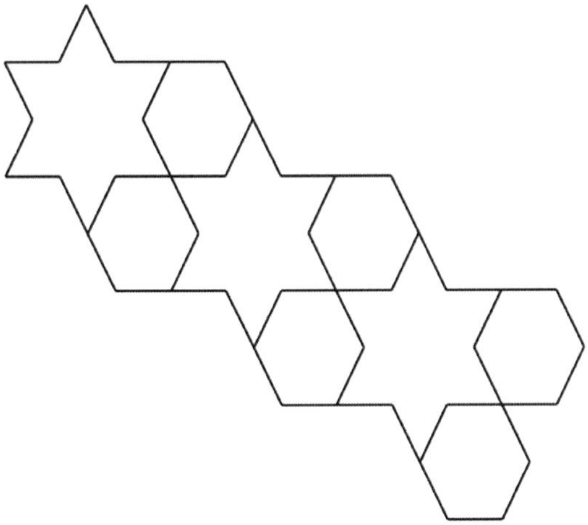

Figure 11.7

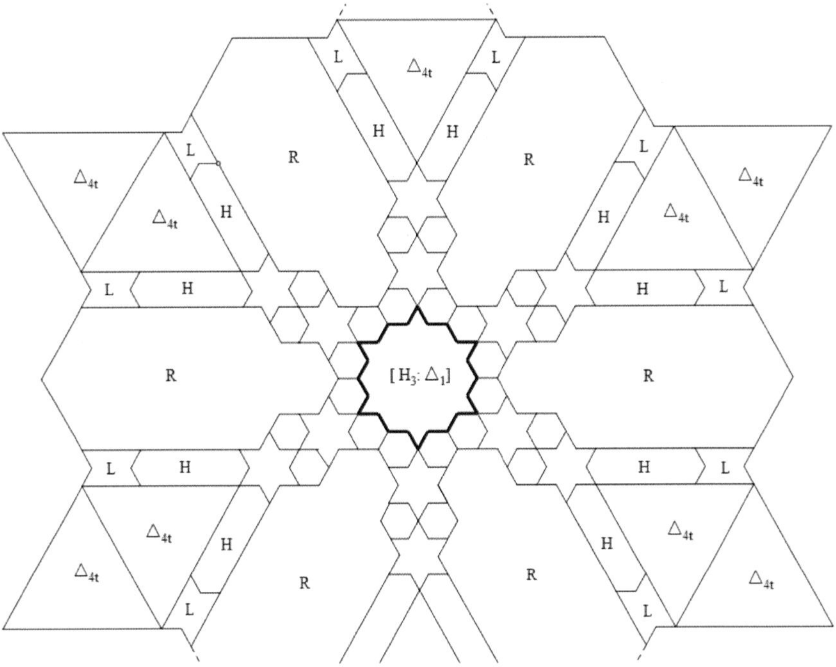

Figure 11.8

In Figure 11.8, the regions labeled "R" are easily tiled with nilpotent parallelograms and triangles. We have only indicated the construction with star chains S^2 emanating from the central $[H_3 : \Delta_1]$. "H" represents the semiregular hexagon H(1,1, 4t-3).

In case t = 0 a composite construction of $[H_m : \Delta_1]$, m = 2s+1; s = 4u+1, proceeds along the same lines with star streams emanating from a central $[H_3 : \Delta_1]$. We leave this construction as an exercise. In case u = 0 the composite construction of $[H_{4t+3} : \Delta_{4t+1}]$ follows a different course as illustrated in Figure 11.9.

Case ii : p = 4t + 1; m = p+2q with q = 4u+3. In this Case, for t≥1, the constructions are similar to those in Case i : $[H_m : \Delta_p]$ can be obtained by appending star chains to $[H_3 : \Delta_1]$. Even for t = 0, a composite construction of $[H_m : \Delta_1]$ can be obtained in this way.

We have proved the following.

Theorem 11.4 *In case p = 4t+1 and m = p+2q with q odd, the irregular star $[H_m : \Delta_p]$ can be constructed as a simple extension of $[H_3 : \Delta_1]$, and thus is composite.* □

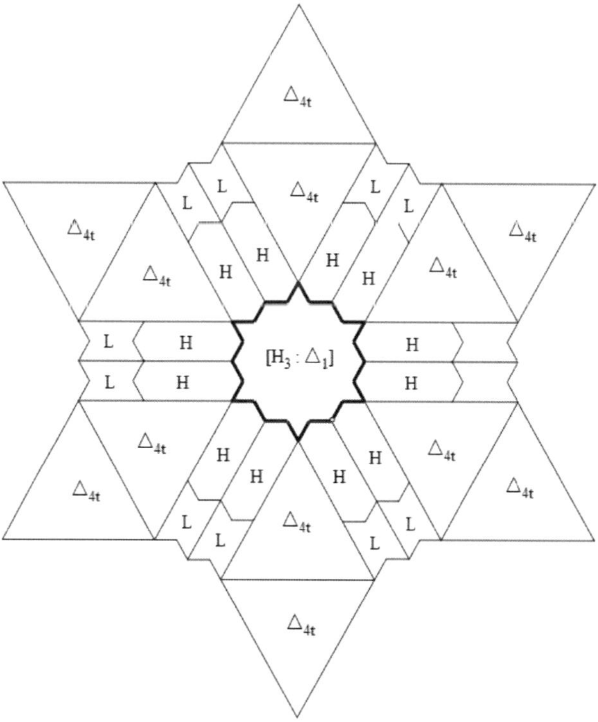

Figure 11.9

Case iii : $p = 4t+3$; $m = p+2q$ with $q = 4u+1$. The irregular stars in this Case can be obtained as simple extensions of members in the set S = {[H_{4k+1} : Δ_3] : k = 1,3,5, ...}. In Figure 11.13 the composite extension [H_5 : Δ_3]→[H_{13} : Δ_3] is llustrated. Further extensions [H_{13} : Δ_3]→[H_{21} : Δ_3]→[H_{29} : Δ_3] are also possible, and thus every member of S can be obtained as a sequence of simple extensions of [H_5 : Δ_3].

The irregular star [H_5 : Δ_3] is a simple extension of the odd shaped polygon, Π_1 indicated with pink boundary in Figure 11.12. We leave it as an exercise to show that Π_1 is a (complex) prime. In Figure 11.12 the simple extension [H_5 : Δ_3]→[H_9 : Δ_7] is also indicated. These observations are summarized in the following result.

Theorem 11.5 *In case $p = 4t+3$; $m = p+2q$ with $q = 4u+1$, the irregular star [H_m : Δ_p] is a simple exension of the complex prime polygon Π_1.*

In Figure 11.11, [H_9 :Δ_3] is constructed as a simple extension of an odd shaped polygon, Π_2 which we show to be prime in the following Theorem.

Theorem 11.6 *The odd shaped polygon Π_2 in Figure 11.11 is a complex prime.*

Proof. In Figure 11.11 there are several pairs of the form (X, -X) since they constitute oppositely valued tiles in a Δ_1 chain. The unpaired values are K, D, H, -C, J, G, B, -F. To show that P is nilpotent it suffices to show that these values sum to 0. The values for the remaining tiles lie in nilpotent figures, as outlined in pink, so these sum to 0. Again considering Δ_1 chains, we obtain the values in the two labeled hexagons near the center of P. The values in each of these hexagons must sum to 0, so we obtain

(1) M + F + C + P = J + G, and
(2) P + H + K + B + D + M = 0.

Eliminating P from (1),(2) yields

(3) J + G - M - F - C + H + K + B + D + M = 0.

Now when M is eliminated (3) reduces to the desired result. □

Case iv : p = 4t+3; m = p+2q with q = 4u+3. The irregular stars in this Case can be obtained as composite extensions of members in the set S = $\{[H_{4k+1} : \Delta_3] : k = 2,4,6, \ldots\}$. In Figure 11.14 the composite extension $\Pi_2 \rightarrow [H_{17} : \Delta_3]$ is illustrated. We could not obtain the further extension $[H_{17} : \Delta_3] \rightarrow [H_{25} : \Delta_3]$ so the genesis of the remaining stars in S remains an open question.

The constructions in Figures 11.6, 11.9 and 11.8 are *simple extensions* of $[H_3 : \Delta_1]$. Figures 11.11, 11.14 illustrate simple extensions of Π_2 and Figure 11.12 illustrates a simple extension of Π_1. In case only simple prime polygons are used in a dissection we call the dissection *simple.*

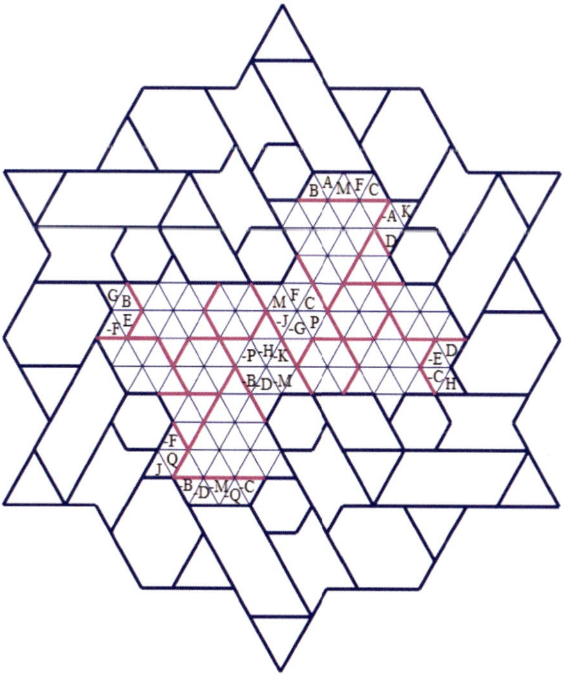

Figure 11.11

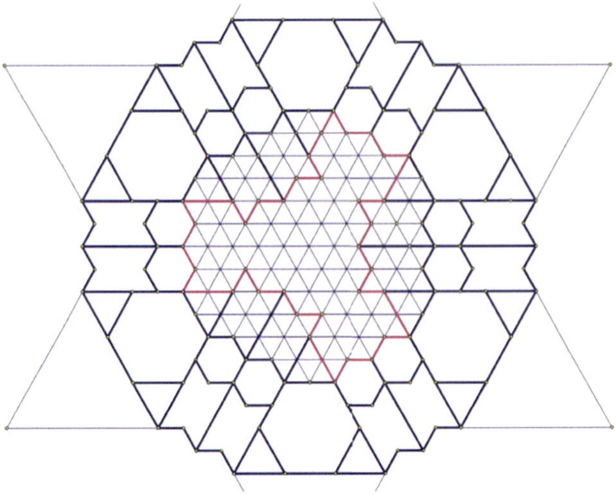

Figure 11.12

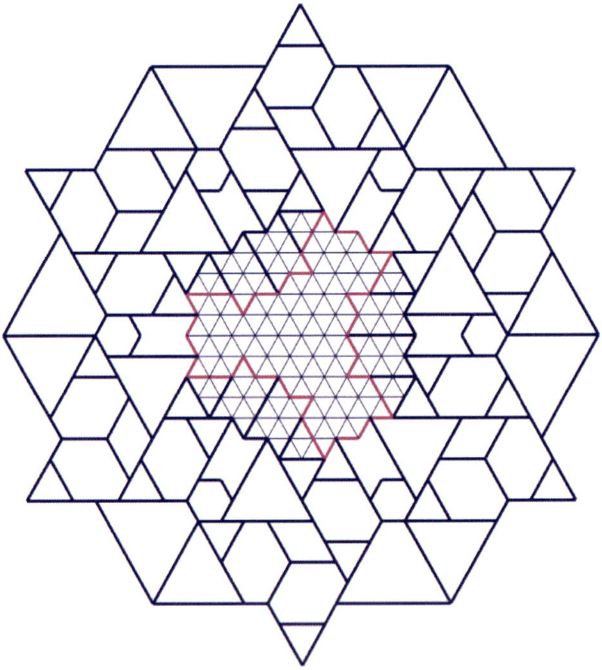

Figure 11.13

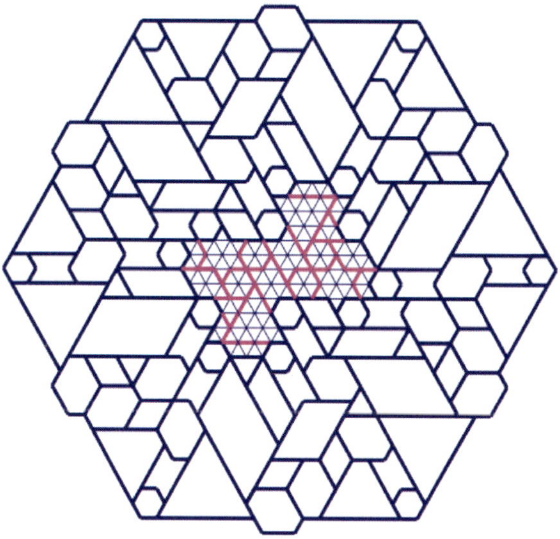

Figure 11.14

11.3 Exercises

(1) Show that the 6-wedge $[H_2 : W^6]$ is prime. (Lemma 11.1)

(2) Find a composite construcion of the irregular star $[H_{11} : \Delta_1]$.

(3) Show that the triangulated polygon Π_1 is prime.

(4) For Case iv, (below Theorem 11.5), how can composite constructions of the remaining members of S be obtained?

(5) All of our constructions of irregular stars are obtained as extensions of the complex prime polygons $[H_3 : \Delta_1]$, Π_1 or Π_2. Can these constructions be obtained as composite extensions of complex prime polygons other than the above mentioned three? Is it possible that some irregular stars have composite constructions in which only the simple prime polygons Δ_2, H_1, L, J_1 and H(2,2,1) are used?

(6) The *pinwheel* $W(H_m : \Delta_p)$, p odd and p<m, is obtained by attaching 6 copies of Δ_p to the sides of H_m so that they are left-

justified on each side. In Figure 11.15(I), the pinwheel $W(H_2 : \Delta_1]$ is illustrated. In 11.15(II), $W(H_3 : \Delta_1]$ is drawn, and 11.15(III) is $W(H_5 : \Delta_3)$. Show that these pinwheels are composite with dissections using only the simple prime polygons Δ_2, H_1, L, J_1 and H(2,2,1). Does every pinwheel have a simple dissection?

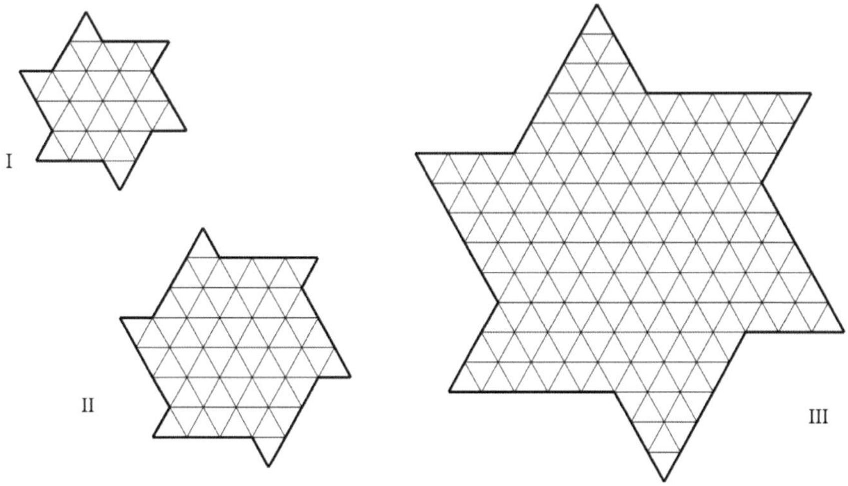

Figure 11.15

(7) Every nilpotent polytriangle so far constructed has even order. Does there exist a nilpotent polytriangle with odd order?

(8) In the most general type of star it should be allowed to attach the triangles Δ_p at any point on the sides of H_m while keeping the order 6 rotational symmetry. The notation $[H_m : s : \Delta_p : q]$ where m = s+p+q will be used to indicate that Δ_p will be attached to each side of H_m with s unit edges to the left and q unit edges to the right. The regular star J_m would be symbolized by $[H_m : 0 : \Delta_m : 0]$ and the pinwheel $W(H_m : \Delta_p)$ would be given by $[H_m : 0 : \Delta_p : m\text{-}p]$ or by $[H_m : m\text{-}p : \Delta_p : 0]$. What has been termed an irregular star $[H_m : \Delta_p]$ would be denoted by $[H_m : s : \Delta_p : s]$ and might better be called a *central irregular star*.

We have now a large collection of stars to investigate. In the problems and Conjectures that follow we consider only the case in which p = 1, and leave the study of stars with p odd and p ≥3 entirely open. Show that $[H_4 : 1 : \Delta_1 : 2]$ is nonnilpotent. In the set of stars $\{[H_m : 1 : \Delta_1 : m\text{-}2]\}$ with $m \geq 4$, which are nonnilpotent and which are composite?

(9) Let $P_t = [H_{4t+1} : s : \Delta_1 : q]$ with s, q even and $s+q = 4t$. Prove or disprove the following Conjecture.

Conjecture 11.7 *If $s = q = 2t$ then P_t has a simple tiling with an order 6 rotational symmetry using only J_1 and H_1. If $s \neq q$ then P_t has a simple tiling with an order 3 rotational symmetry using only J_1, H_1, Δ_2 and L. The pinwheel $[H_4 : 0 : \Delta_1 : 4]$ has a unique tiling using only H_1, Δ_2 and L.*

In Figure 11.16(I) the stipulated tiling of P_1 with s=q=2 illustrated. The pinwheel $[H_4 : 0 : \Delta_1 : 4]$ has a unique tiling using only H_1, Δ_2 and L as in Figure 11.16(II).

Figure 11.16

(10) Let $Q_t = [H_{4t+3} : s : \Delta_1 : q]$ with s, q even and $s+q = 4t+2$. Prove or disprove the following Conjecture.

Conjecture 11.8 *Q_t has a dissection with a rotational symmetry of order 3 which uses only H_1, Δ_2, L and H(2,2,1) as components.*

Can a dissection of Q_t be achieved as an extension of a pinwheel?

(11) Let Q_t be defined as in problem (10) but with s, q both *odd.* Prove or disprove the following Conjecture.

Conjecture 11.9 *If* $s+q = 4t+2$ *then* Q_t *can be tiled as a simple extension of the prime central irregular polygon* $[H_3 : \Delta_1] = [H_3 : 1 : \Delta_1 : 1]$.

(12) Let P_t be defined as in problem (9) but with s, q both *odd.* Prove or disprove the following Conjecture.

Conjecture 11.10 *If* $s+q = 4t$ *then* P_t *is nonnilpotent.*

(13) Let $R_t = [H_{4t} : s : \Delta_1 : q]$ with $s+q = 4t-1$, and let $U_t = [H_{4t+2} : s : \Delta_1 : q]$ with $s+q = 4t+1$. Prove or disprove the following Conjecture.

Conjecture 11.11 R_t *is nonnilpotent and* U_t *is composite.*

In Figure 11.17(I) a composite tiling of U_1 with s=1 and q=4 is illustrated. In Figure 11.17(II) the polytriangle R_1 is drawn with s =1 and q=2. Show that R_1 is nonnilpotent.

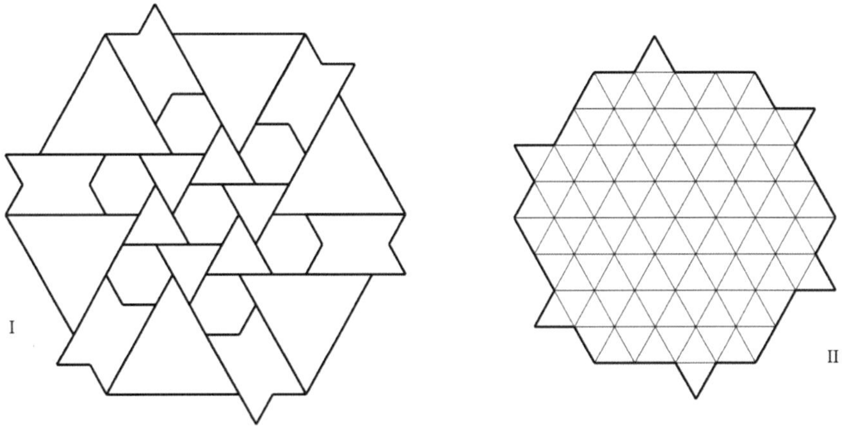

Figure 11.17

(14) In Figure 11.18 the star $S = [H_7 : 2 : \Delta_1 :4]$ is illustrated with a dissection consisting of a simple extension of a complex polygon Π_3. Show that Π_3 is nilpotent. Is Π_3 also prime? Find also a simple dissection of S consistent with Conjecture 11.18.

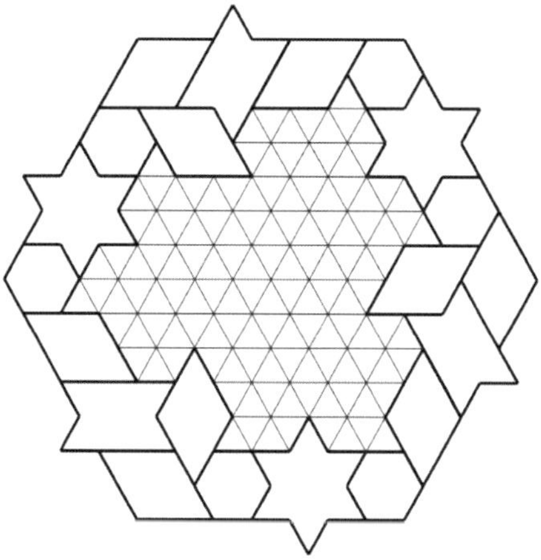

Figure 11.18

(15) Triangular extensions also yield some interesting nilpotent polytriangles. These will be denoted by $[\Delta_m : s : \Delta_p : q]$ where m is odd and m = s+p+q. The triangle Δ_p is attached to each side of Δ_m with s units to the left and q units to the right. We will just consider examples with p = 1. In Figure 11.19(I) the polytriangle $[\Delta_7 : 5 : \Delta_1 : 1]$ is illustrated.

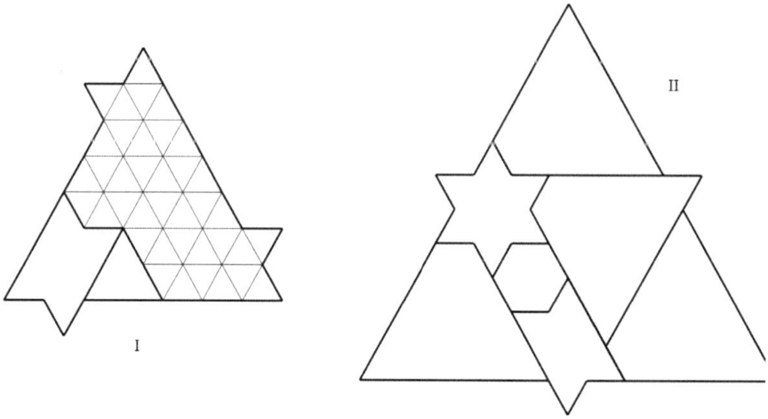

Figure 11.19

Show that $[\Delta_7 : 5 : \Delta_1 : 1]$ is an extension of the indicated odd-shaped polytriangle Λ_1, and that Λ_1 is a complex prime. Note that since m is odd $\nu(\Delta_m)$ is the sum of values assigned to the three corner unit triangles. The appended Δ_1 triangles are placed so that their value is the negative of the corner value on the opposite side of Δ_m. In this way the nilpotent polytriangle $[\Delta_{11} : 5 : \Delta_1 : 5]$ in Figure 11.19(II) was obtained. A simple dissection is also indicated. The Δ_1 attachments can be placed in different positions even while retaining a rotational symmetry of order 3. Consider, for example, the $[\Delta_{11} : 1 : \Delta_1 : 9]$ in Figure 11.20.

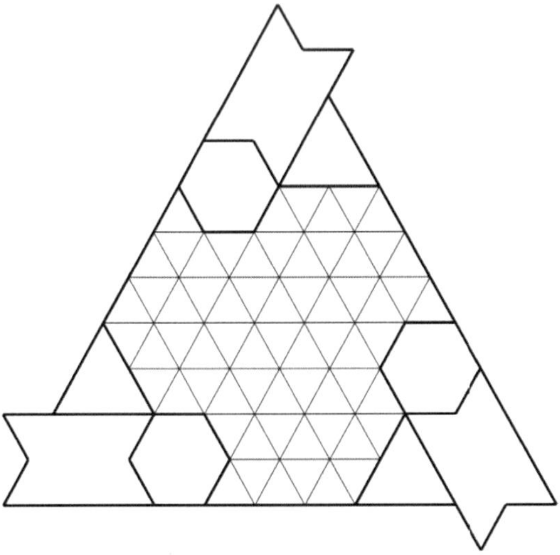

Figure 11.20

Show that the odd shaped polygon Λ_2 indicated in Figure 11.20 is prime and thus $[\Delta_{11} : 1 : \Delta_1 : 9]$ is composite. Which of the many nilpotent polytriangles of this kind with order 3 rotational symmetry are (i) simple extensions of complex prime polygons or (ii) have simple dissections? For the ones which are extensions of complex primes, can they all be constructed as simple extensions of Λ_1 or Λ_2?

(16) In Figure 11.21 a new tiling of $[H_9 : \Delta_3]$ is illustrated which features a simple extension of the complex polytriangle Π_4, outlined in brown. Show that Π_4 is nilpotent. Is Π_4 also prime ?

(17) Suppose the polytriangle P is nilpotent and P is a simple extension of a polytriangle Q. Suppose also that every locally

nilpotent assignment for Q extends to a locally nilpotent assignment for P. Then it must be that Q is itself nilpotent. Use this method to show that Λ_1, (in problem (15)), is nilpotent. Use this method to show that Π_4 in problem (16) is nilpotent. What properties must the boundary of Q have to insure that every locally nilpotent assignment for Q extends to a locally nilpotent assignment for P?

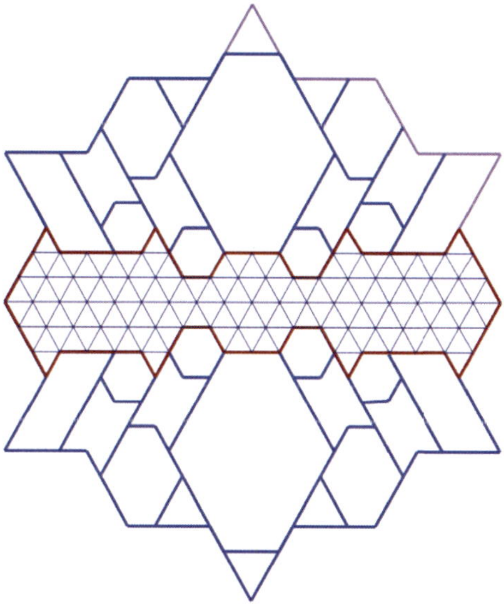

Figure 11.21

(18) In Figure 11.12 the irregular star $[H_5 : \Delta_3]$ is depicted as a simple extension of the complex prime Π_1. In Figure 11.22 a decomposition of $[H_5 : \Delta_3]$ consisting of 6 H_1 hexagons, 6 Δ_2 triangles and a polytriangle Π_5 with brown boundary. Show that Π_5 is prime. If the indicated central H_1 hexagon, (in pink), of Π_5 is removed, show that the resulting polytriangle with a hole remains nilpotent, but it cannot in general be completed to obtain a nilpotent Π_5.

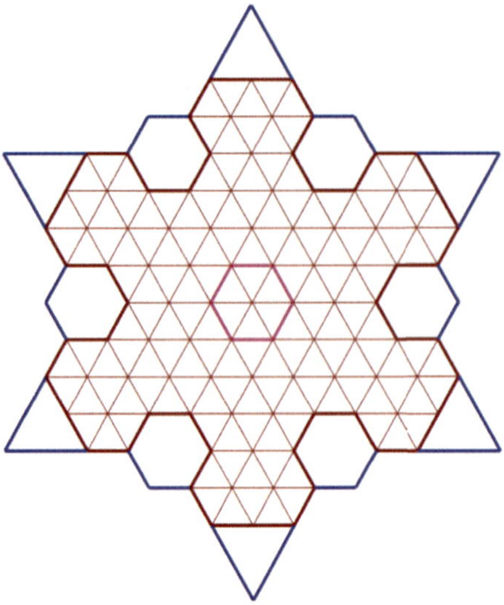

Figure 11.22

CHAPTER 12

The Game of Triangles

This game is similar to the game of rhombs (Chapter 9) but is played with equilateral (unit) triangles as tiles instead of rhombs. Each tile is marked with a value in the range {0, ... , 9} and scoring configurations consist of the prime configurations Δ_2, H_1, J_1 and H(2,2,1) whose value is a multiple of 5. There are a total of 80 tiles with 8 of each denomination. All tiles are placed face down in the *boneyard* and each player selects 8 tiles for his/her hand and places them on edge so their values can be seen.

Suppose there are two players A and B. Player A places one tile from his hand onto the table. No score is possible so player A completes his hand by drawing a single tile from the boneyard to replenish his hand to a total of 8 tiles. Player B then places one of her tiles adjacent to the tile played by A. Still no score is possible, so B replenishes her hand and play reverts to A. Player A places one of his tiles adjacent to an already played tile; still no score is possible, so A replenishes his hand and completes his turn.

Now for the first time player B has a possibility of making a score by completing a Δ_2 with value a multiple of 5. *As in the Game of Rhombs no tile can be played in a position which creates one of the five basic scoring configurations but which does not yield a value which is a multiple of 5.* Thus in Figure 12.1(I), if Player B has a tile with value 1 or 6 then it can be played in position X to obtain a score of 15 or 20 respectively, otherwise player B must place her tile in one of the nonscoring positions marked Y. In case B can play in position X, then in the same turn B must play an additional tile before completing her turn and replenishing her hand.

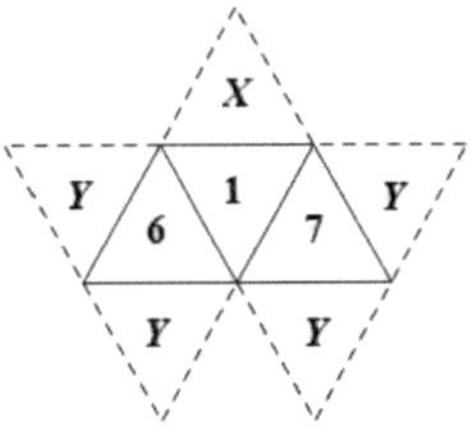

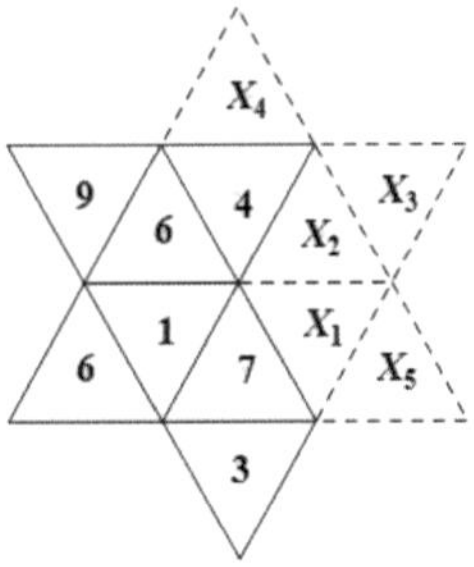

Figure 12.1

In Figure 12.1(II) suppose it is player A's turn and A has tiles with values 4, 8, 9, 2, 1 in his hand. In one turn A can play his tiles in the sequence $X_1 = 4$; $X_2 = 8$; $X_3 = 9$; $X_4 = 2$; $X_5 = 1$. Playing $X_1 = 4$ completes a Δ_2 and yields a score of 15. Playing $X_2 = 8$ completes an H_1 hexagon and yields a score of 30. Playing $X_3 = 9$ completes a Δ_2 scoring 25. Playing $X_4 = 2$ completes a Δ_2 with a score of 20. Finally, playing $X_5 = 1$ completes the star J_1 with a score of 60. (Note that the tiles occupying opposite points of J_1 must sum to a multiple of 5.) The total score for A is 150. *A must complete his turn by playing a nonscoring tile adjacent to a tile in the central polytriangle, and then drawing 6 tiles from the boneyard.*

A configuration like the one illustrated in Figure 12.2 occurs quite often and must be considered carefully. The location marked X simultaneously completes two potential scoring configurations: a Δ_2 and an H_1. In Figure 12.2(I) X = 7 completes the Δ_2 with a score of 20, and completes H_1 with a score of 25 yielding a total score of 45 for the play. In Figure 12.2(II) only one value is changed but this makes any play at position X impossible. The reason is that X = 3 or X = 8 is required to complete the Δ_2 and X = 2 or X = 7 is required to complete the H_1 to a scoring configuration. Thus no one value for X can complete both scoring configurations. The consequence is that no tile can be placed at X. Position X is called a *dead location* and should be marked as such, for example, with a penny. The configuration in 12.2(II) occurs much more frequently than the scoring configuration in 12.2(I), and players should be constantly on the lookout for dead locations.

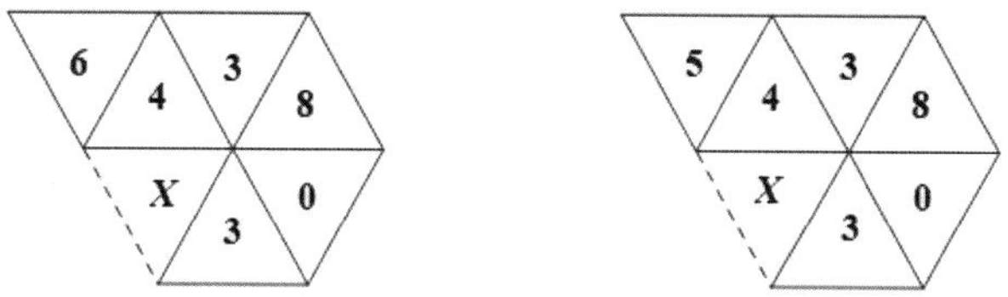

I II

Figure 12.2

Play alternates between A and B until all 80 tiles have been played. If one player has played all of his/her tiles and the boneyard is exhausted, then that player must pass. The other player then plays all of his/her tiles in sequence until all tiles have been played. There is no bonus and the game does not end when a player plays every tile in his/her hand. When all tiles have been played the *set* is complete and the scores tallied. The player with the highest score wins the set. Any number of sets can be played.

Playing time is considerably reduced by starting with 9 instead of 8 tiles, and replenishing to 9 tiles at the end of each turn. Three or four players is possible with a reduction to 7 tiles per hand. This game, as well as the Game of Rhombs can easily be constructed using mat board, an Exacto knife, straight edge and compass. 1.5 inches per side of each tile works very well.

Besides the 5 basic scoring configurations, players may wish to include Δ_4 and H_2. The triangle Δ_4 can occur quite frequently, whereas the hexagon H_2 occurs only rarely. In our opinion nilpotent parallelograms or trapezoids and even the link L should not be included as scoring configurations even though they occur frequently. Scoring can become quite tedious if these configurations are included.

INDEX

SYMBOL LIST

REFERENCES

[1] Fletcher III, R.R. Magic Polygons, Westwood Books Publishing (2022)

[2] Fletcher III, R.R. Polyparallelogram Puzzles and Tiling Problems, Westwood Books Publishing (2022)

[3] Akiyama J., Kano M. Factors and Factorizations of Graphs, Springer-Verlag, (2011)

[4] Ore O. The four Color Problem, Academic Press, New York (1967)

[5] Grunbaum B., Shephard G.C. Tilings & Patterns, Dover Publications, (2016)

Made in United States
Orlando, FL
04 November 2026

90207137R00098